Plant-derived Pharmaceuticals

Principles and Applications for Developing Countries

CABI Biotechnology Series

Biotechnology, in particular the use of transgenic organisms, has a wide range of applications including agriculture, forestry, food and health. There is evidence that biotechnology could make a major impact on producing plants and animals that are able to resist stresses and diseases, thereby increasing food security. There is also potential to produce pharmaceuticals in plants through biotechnology, and provide foods that are nutritionally enhanced. Genetically modified organisms can also be used in cleaning up pollution and contamination. However, the application of biotechnology has raised concerns about biosafety, and it is vital to ensure that genetically modified organisms do not pose new risks to the environment or health. To understand the full potential of biotechnology and the issues that relate to it, scientists need access to information that not only provides an overview of and background to the field, but also keeps them up to date with the latest research findings.

This series, which extends the scope of CABI's successful Biotechnology in Agriculture series, addresses all topics relating to biotechnology including transgenic organisms, molecular analysis techniques, molecular pharming, *in vitro* culture, public opinion, economics, development and biosafety. Aimed at researchers, upper-level students and policy makers, titles in the series provide international coverage of topics related to biotechnology, including both a synthesis of facts and discussions of future research perspectives and possible solutions.

Titles Available

1. Animal Nutrition with Transgenic Plants
   Edited by G. Flachowsky
2. Plant-derived Pharmaceuticals: Principles and Applications for Developing Countries
   Edited by K.L. Hefferon

Plant-derived Pharmaceuticals

Principles and Applications for Developing Countries

———————————

Edited by

Kathleen L. Hefferon

Cornell University
Ithaca, New York, USA

www.cabi.org

CABI is a trading name of CAB International

CABI	CABI
Nosworthy Way	38 Chauncy Street
Wallingford	Suite 1002
Oxfordshire OX10 8DE	Boston, MA 02111
UK	USA
Tel: +44 (0)1491 832111	T: +1 800 552 3083 (toll free)
Fax: +44 (0)1491 833508	E-mail: cabi-nao@cabi.org
E-mail: info@cabi.org	
Website: www.cabi.org	

A catalogue record for this book is available from the British Library, London, UK.

Library of Congress Cataloging-in-Publication Data

Plant-derived pharmaceuticals : principles and applications for developing countries / editor: Kathleen L. Hefferon, Cornell University, Ithaca, NY.
    pages cm. -- (CABI biotechnology series ; 2)
 Includes bibliographical references and index.
 ISBN 978-1-78064-343-4 (alk. paper)
 1. Materia medica, Vegetable--Developing countries. 2. Medicinal plants--Biotechnology--Developing countries. 3. Pharmaceutical biotechnology--Developing countries. 4. Vaccines--Biotechnology--Developing countries. 5. Monoclonal antibodies--Biotechnology--Developing countries. I. Hefferon, Kathleen L. II. Series: CABI biotechnology series ; 2.

 RS164.P7273 2014
 615.3'21091724--dc23

                                 2014006676

ISBN-13: 978 1 78064 343 4

Commissioning editor: David Hemming
Editorial assistant: Alexandra Lainsbury
Production editor: Tracy Head

Typeset by Columns Design XML Ltd, Reading, UK.
Printed and bound by CPI Group (UK) Ltd, Croydon, CR0 4YY.

Contents

Contributors

José Francisco Castillo Esparza, Centro de Investigación y de Estudios Avanzados (CINVESTAV), Unidad Irapuato, Guanajuato, México. E-mail: jesparza@ira.cinvestav.mx

Qiang Chen, Center for Infectious Diseases and Vaccinology, Biodesign Institute and School of Life Sciences at Arizona State University, USA. E-mail: Qiang.Chen.4@asu.edu

Marcello Donini, Dipartimento BAS, Sezione Genetica e Genomica Vegetale, ENEA, C.R. Casaccia, via Anguillarese 301, 00123, ENEA, Italy. E-mail: Marcello.donini@ enea.it

Pascal M.W. Drake, Hotung Molecular Immunology Unit, Institute for Infection and Immunity, St George's University of London, Cranmer Terrace, London, SW17 0RE. E-mail: pdrake@sgul.ac.uk

Miguel Angel Gómez Lim, Centro de Investigación y de Estudios Avanzados (CINVESTAV), Unidad Irapuato, Guanajuato, México. E-mail: mgomez@ira.cinvestav.mx

Sonia Gutiérrez, Southern Crop Protection and Food Research Centre, Agriculture and Agri-Food Canada.

Kathleen L. Hefferon, Cornell University, USA. E-mail: klh22@cornell.edu

Inga I. Hitzeroth, Rondebosch UCT University of Cape Town, South Africa. E-mail: Inga. Hitzeroth@uct.ac.za

Huafang Lai, Center for Infectious Diseases and Vaccinology, Biodesign Institute and School of Life Sciences at Arizona State University, USA. E-mail: Huafang.Lai@asu.edu

Elizabeth Loza-Rubio, National Center of Veterinary Microbiology (CENID-Microbiologia), INIFAP, Mexico. E-mail: loza.elizabeth@inifap.gob.mx

Carla Marusic, Dipartimento BAS, Sezione Genetica e Genomica Vegetale, ENEA, C.R. Casaccia, via Anguillarese 301, 00123, ENEA, Italy. E-mail: carla.marusic@enea.it

Karen A. McDonald, Global HealthShare Initiative, Department of Chemical Engineering and Materials Science, University of California, Davis, One Shields Avenue, Davis, CA 95616, USA. E-mail: kamcdonald@ucdavis.edu

Rima Menassa, Southern Crop Protection and Food Research Centre, Agriculture and Agri-Food Canada. E-mail: Rima.Menassa@agr.gc.ca

Ann E. Meyers, Rondebosch UCT University of Cape Town, South Africa. E-mail: Ann. Meyers@uct.ac.za

García Alberto Monroy, Centro de Investigación y de Estudios Avanzados (CINVESTAV), Unidad Irapuato, Guanajuato, México. E-mail: agarcia@ira.cinvestav.mx

Somen Nandi, Global HealthShare Initiative, Department of Molecular and Cellular Biology, University of California, Davis, One Shields Avenue, Davis, CA 95616, USA. E-mail: snandi@ucdavis.edu

Edith Rojas-Anaya, National Center of Veterinary Microbiology (CENID-Microbiologia), INIFAP, Mexico. E-mail: rojas.edith@inifap.gob.mx

Edward P. Rybicki, Rondebosch UCT University of Cape Town, South Africa. E-mail: Edward.Rybicki@uct.ac.za

Sonia Sadone, Centre for Infection, St George's University of London, UK. E-mail: Ssadone@sgul.ac.uk

Harry Thangaraj, Centre for Infection, St George's University of London, UK. E-mail: hthangaraj@sgul.ac.uk

1 Introduction, and the Promise of a Plant-derived Vaccine for Hepatitis B Virus

Kathleen L. Hefferon*

Cornell University, Ithaca, New York

1.1 Introduction

It is a disturbing fact that in the developing world today, routinely preventable infectious diseases remain a leading source of childhood mortality (Fig. 1.1). Fundamental difficulties in accessing vaccines and other biopharmaceuticals offer one explanation for these statistics. In developing countries, the high cost and limited accessibility of vaccines and other therapeutic proteins for remote, impoverished regions continues to prevent essential help to children in need. More than 20 years ago, the Children's Vaccine Initiative, an assembly of philanthropic groups, joined forces with the World Health Organization to tackle this problem by striving to develop efficacious vaccines that can be safely delivered orally and in a cost-effective manner, without the requirement for refrigeration or significant medical infrastructure (United Nations, 2010; Hefferon, 2013). The utilization of plants as production platforms as well as vehicles for oral delivery of vaccine proteins and other biopharmaceuticals represented one of the more promising innovations that arose from this initiative.

There is no doubt that plant-derived vaccines have the potential to improve the lives of many in the developing world. This technology can be used to curtail the spread of infectious diseases such as hepatitis B virus (HBV), human immunodeficiency virus (HIV) and malaria. Plant-derived vaccines could in the future be employed to prevent infant diarrhoeal diseases caused by rotavirus, cholera and enterotoxigenic *Escherichia coli*. In addition to this, orphan diseases such as dengue fever and rabies, which tend to be associated with poor financial involvement by the West, could also be tackled with renewed vigour (Streatfield, 2006; Bethony *et al.*, 2011; Penney *et al.*, 2011). Recently, government funding has been provided for the development of plant-derived vaccines against global pandemics such as influenza H1N1 virus, as well as potential biological warfare agents such as smallpox and anthrax. New vaccines are required, which are both inexpensive and can be up-scaled en masse rapidly, and the use of plants as a production platform provides a suitable match for this purpose (McCormick *et al.*, 2008; Rigano *et al.*, 2009; Chichester *et al.*, 2012, 2013).

* E-mail: klh22@cornell.edu

© CAB International 2014. *Plant-derived Pharmaceuticals: Principles and Applications for Developing Countries* (ed. K.L. Hefferon)

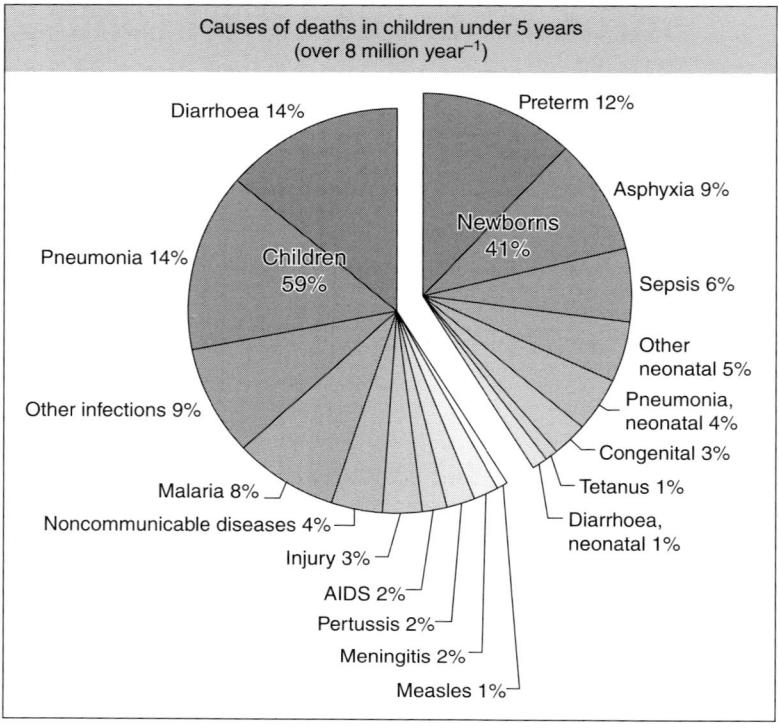

Fig. 1.1. Causes of death in children under 5 years (WHO, 2013).

Plants have a number of advantages over conventional protein expression systems, including mammalian and bacterial cultures. For example, plant cells are not able to be infected by the same pathogenic agents as mammalian cells, thus they are safer to use. Plant cells also are capable of undergoing post-translational modifications that closely resemble their mammalian counterparts, making them more attractive to use than bacterial-based expression systems. Plants are inexpensive to grow, and vaccines produced from them can be easily purified or require only partial purification prior to oral administration. The fact that many plant tissues can be orally ingested means that the vaccine proteins will be directly presented to the mucosal immune system via the gut-associated lymphoid tissue (GALT) (Paul and Ma, 2010). A selective advantage lies in the fact that in several cases, consumption of vaccines through edible plant tissue has provided a stronger mucosal response than conventional oral vaccines; this may be due to the fact that the plant tissue is better able to protect the vaccine antigen as it passes through the harsh environment of the gastrointestinal tract.

Plants can be utilized to produce vaccines within multiple formats, ranging from the generation of stable transgenic plants via nuclear or chloroplast transformation, to transient infection via agroinoculation or through the use of plant virus expression vectors. Plants can be grown in the field, in the greenhouse, or under tissue culture conditions. Vaccine proteins can be expressed under developmental or tissue-specific promoters, so that the vaccine protein in question can be expressed only in the seed of maize, the ripened fruit of tomato, or the root of carrot, as examples. The decision over which plant expression platform should be used to express a specific vaccine protein thus should involve the consideration over which plant species and expression system is the best fit, be it cell line or whole plant, stable or transient expression.

This book describes the most recent research involving the generation of vaccines and other therapeutic proteins for the specific purpose of combating infectious diseases which predominate in developing countries. Each chapter of the book will be presented by a world authority in the field, and will include an up-to-date review of the recent literature on each subject. Topics within the book encompass the current state-of-the-art technologies used for developing vaccines and other therapeutic proteins from plants. A selection of case studies will also be provided, detailing the use of plant-made vaccines to protect against some of the most pervasive infectious diseases found in the Third World today. The book concludes with a discussion of the future of plant-made pharmaceuticals for developing countries, and whether plant-made vaccines can realistically make a difference for the world's poor.

One of the first plant-derived therapeutic agents generated with developing countries in mind is a vaccine to prevent infection by HBV. It seems fitting, therefore, that the first chapter in this book should include a description of some of the history of this vaccine and where things lie at present with respect to its future use. HBV is also covered in other chapters of this book.

1.2 Hepatitis B Virus

It is striking to realize that approximately one-third of the human population is infected with HBV, and the number of chronic HBV carriers continues to increase each year. A causative agent of chronic liver disease, HBV has brought about considerable economic loss throughout the world due to human mortality and morbidity (Fig. 1.2). During infection, HBV produces large amounts of an excess virion surface protein known as hepatitis B surface antigen (HBsAg), which can circulate throughout the bloodstream of carriers as 22 nm diameter non-infectious virus particles. These particles themselves, upon parenteral administration, can elicit a robust immune response. The first vaccine against HBV had to be generated from the plasma of infected individuals, since the virus could not be successfully propagated in tissue culture. Subunit HBV vaccines based on HBsAg have been generated in either yeast or mammalian cell culture systems, and were introduced into international vaccine programmes in the 1980s. However, in developing countries, where HBV infection continues to predominate, insufficient medical infrastructure prevails and the cost of providing this vaccine en masse makes its use unfeasible. The generation of a plant-derived version of

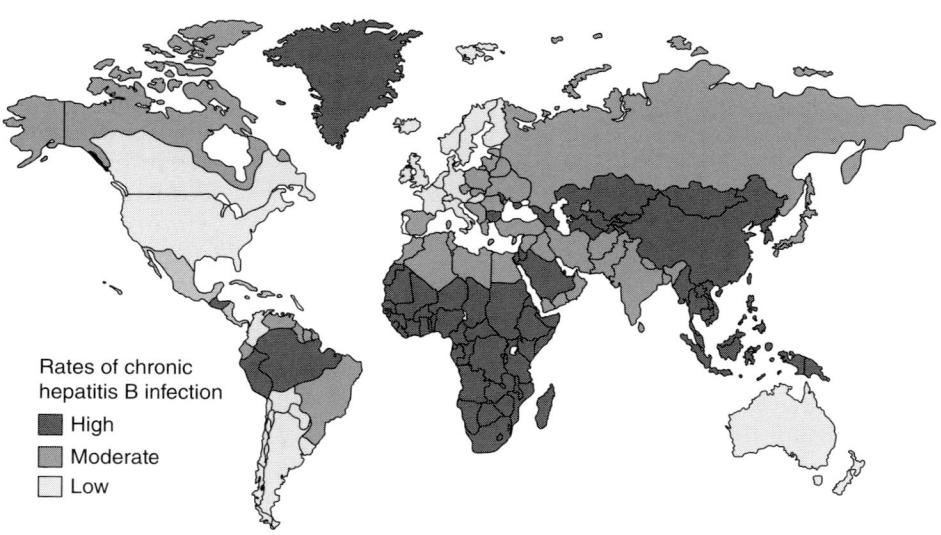

Fig. 1.2. Worldwide rates of chronic hepatitis B virus (Centers for Disease Control and Prevention, http://www.cdc.gov/hepatitis/populations/api.htm).

the HBsAg vaccine would therefore provide a means by which to combat the disease on a global scale and in a cost-effective manner (Shchelkunov and Shchelkunova, 2010).

This plant-based vaccination strategy for HBV has been explored by a number of research groups. Initially, model transgenic plant systems such as potato, tomato and tobacco were used as vaccine expression platforms. Studies which were based upon providing more efficient expression led to the use of plant tissues or crude extracts of plants to conduct animal trials (Guan *et al.*, 2010). For example, one of the earliest studies was performed in BALB/c mice. An immune response was readily demonstrated in those mice which were parentally administered a crude transgenic plant extract. These mice were induced to produce an immune response to the plant-derived vaccine that was similar in size and range to that induced by the commercially available yeast-derived vaccine. Both humoral and T cell mediated responses were elicited in a manner that mimicked the commercially available yeast-derived vaccine. An innovative version of HBsAg was also designed, which has a modified N-terminus; this version could be used to present both T cell and B cell epitopes. Modification of the antigen did not negatively impact its ability to form virus-like particles, indicating that this modified version of the antigen can be used to elicit a multivalent response.

Potato tubers expressing the surface antigen of HBV have also been fed to individual human volunteers (Richter *et al.*, 2000). The patients had been previously vaccinated for HBV and were fed 100 g raw potato tubers on days 0, 14 and 28; the vast majority were able to elicit a strong IgG titre in their blood serum. This was the first proof-of-concept for the production of an immune response in humans based upon the oral consumption of a vaccine antigen produced in a food crop plant.

Many other investigations regarding the immune response to plant-derived vaccines against HBV have also been conducted. In a series of studies by the research group of M. Sala, a bivalent vaccine composed of recombinant HIV-1/HBV virus-like particles (VLPs) were expressed in *Nicotiana tabacum* and *Arabidopsis thaliana* plants (Greco *et al.*, 2007). In this instance, an HIV-1 polyepitope was expressed in the form of HBV virus-like particles. When tissue from these transgenic plants expressing the VLPs were ingested by humanized HBV inoculated mice as a boost to DNA priming, an anti-HIV-1 specific CD8+ T cell activation was elicited, which could be detected in mesenteric lymph nodes (Guetard *et al.*, 2008). The work is significant because both HIV-1 and HBV are transmitted in a similar fashion and infection by HBV occurs in up to 32% of HIV-1 cases.

The surface antigen of HBV has also been expressed in rice under regulation of the seed-specific promoter Glub-4. The *SS1* gene used in this study comprises a fusion protein consisting of amino acids 21–47 of the hepatocyte receptor-binding pre-surface 1 region (pre-S1) fused to the truncated C-terminus of the major HBV surface (S) protein (Qian *et al.*, 2008). Antibodies that can be elicited from this pre-S1 region could then block against HBV binding to hepatocytes. This HBV fusion product was expressed at levels as great as 31.5 ng g^{-1} dry weight in rice seed and could be detected by both an anti-S protein antibody and an anti-pre-S1 antibody. BALB/c mice immunized with the recombinant SS1 protein were able to generate immunological responses against both the S and pre-S1 epitopes.

Youm *et al.* (2010) used a variety of orally administered doses of potato-derived HBsAg to determine their effects on the mouse immune response. HBV-specific IgA and IgG responses were observed even at the lowest dose of 0.02 μg HBsAg. The precise mechanisms behind the IgG immune response were investigated further through the analysis of IgG subclasses. The authors found that the level of IgA was greater in mice that were fed plant-derived rather than yeast-derived HBsAg, suggesting that the former is more stable in the gut. The authors were also able to observe M cell uptake of plant-derived HBsAg in the Peyer's patches, indicating that the vaccine is able to successfully enter the desired target tissue and elicit an immune response.

Regardless of the success of these animal and human clinical trials, the fact remains that many plants consumed as foods, such as potato, cannot be ingested raw. Cooking is not a feasible option because this would most likely degrade the immunogen expressed in the plant tissue. To circumvent this problem, transgenic tomato plants have also been used as an expression platform for HBsAg. Lou *et al.* (2007) used different construct designs to optimize levels of HBV antigen in tomato. These included providing a more 'plant friendly' codon usage, an amino acid sequence KDEL endoplasmic reticulum retention signal, and a tomato fruit-specific promoter known as 2A11. In addition to this, the tobacco pathogenesis-related protein S signal peptide was also fused to the 5′ terminus of the open reading frame of the construct. All of these factors contributed to higher and more stable accumulation of HBsAg in the tomato plants. Using these combined approaches, the researchers were able to increase the expression of HBsAg to 0.02% of the soluble protein. The authors also found that the accumulation of HBsAg was 65- to 171-fold larger in mature fruit than in small or medium-sized fruits and leaf tissues. Further investigation using electron microscopy indicated that HBsAg expressed in the fruit of tomato can assemble into capsomeres and virus-like particles. HBsAg has also been expressed in lettuce that is adapted to warmer climates, such as Brazil, through the generation of transgenic plants via co-cultivation of cotyledons with *Agrobacterium* harbouring the gene of interest (Marcondes and Hansen, 2008). Lettuce represents another plant species that can be consumed raw, and is thus a more feasible expression platform for oral vaccines.

Transgenic plants have also been generated that express the HBV core antigen. These have been demonstrated to self-assemble into nucleocapsid particles that contain both viral RNA and polymerase molecules. These particles were shown to have an immunostimulatory effect when combined with HBsAg but cannot alone protect against HBV infection.

Recently, a so-called 'third generation' tri-component vaccine was developed in tobacco and lettuce; this contained the small S-HBsAg in addition to the other envelope proteins of HBV, including the medium (M-HBsAg) and large (L-HBsAg) surface antigens, all of which are expressed from the same open reading frame (Pniewski *et al.*, 2011, 2012). The antigens displayed a common S domain and characteristic domains pre-S2 and pre-S1 and were assembled into virus-like particles (VLPs), which are both stable and highly immunogenic. The S fragment also anchors M and L antigens within the virus membrane, and as a result, expression of all of the HBV surface antigens enhance immunogenicity by eliciting a larger spectrum of antiviral antibodies. The resulting transgenic plants produced these three HBV antigens in the form of self-assembled VLPs or analogous aggregates. The stability of these particles as multimeric aggregates was investigated by freeze-drying lyophilized plant tissue. Lettuce was used as the target species in these studies, as it is consumed raw and can easily be lyophilized and freeze-dried to increase vaccine concentration. Lettuce is also an attractive crop to produce an orally administered antigen, as it is naturally free of alkaloids and other harmful compounds. The efficiency of lyophilization and storage was found to depend on the initial antigen content in plant tissue, yet M-HBsAg appeared to be approximately 1.5–2 times more stable than L-HBsAg. The tissue can be lyophilized into a powder, which can be formed into tablets or capsules. The authors used these experiments to further determine optimal conditions for storage, dosage and administration of vaccine prior to clinical trials, to identify further its potential for use as a prototype oral vaccine. The antigen was demonstrated to be stable for at least 1 year when stored at room temperature (Pniewski *et al.*, 2012).

While these results constitute a promising step forward into making plant-derived HBV vaccines a reality for developing countries, more thought will be required to see this vaccine implemented as part of an effective immunization procedure in the near future. One possibility that appears to be economically feasible is the use of a two component prime-boost plant-derived vaccine, based on parenteral and oral immunization, respectively (Pniewski, 2013). Further studies will help to elucidate the potential success of this approach. Without question, the history and continuing research involved in developing a plant-derived vaccine for HBV represents a prominent case in point with respect to the field of plant-made pharmaceuticals. Other vaccines have their own individual stories, as demonstrated in other chapters. It is with this first example in mind that the reader is invited to explore other models of vaccine development offered in this book, and their feasibility for providing relief to developing countries.

References

Bethony, J.M., Cole, R.N., Guo, X., Kamhawi, S., Lightowlers, M.W., Loukas, A., Petri, W., Reed, S., Valenzuela, J.G. and Hotez, P.J. (2011) Vaccines to combat the neglected tropical diseases. *Immunology Reviews* 239(1), 237–270.

Chichester, J.A., Jones, M.R., Green, B.J., Stow, M., Miao, F., Moonsammy, G., Streatfield, S.J. and Yusibov, V. (2012) Safety and immunogenicity of a plant-produced recombinant hemagglutinin-based influenza vaccine (HAI-05) derived from A/Indonesia/05/2005 (H5N1) influenza virus: a phase 1 randomized, double-blind, placebo-controlled, dose-escalation study in healthy adults. *Viruses* 4, 3227–3244.

Chichester, J.A., Manceva, S.D., Rhee, A., Coffin, M.V., Musiychuk, K., Mett, V., Shamloul, M., Norikane, J., Streatfield, S.J. and Yusibov, V. (2013) A plant-produced protective antigen vaccine confers protection in rabbits against a lethal aerosolized challenge with *Bacillus anthracis* Ames spores. *Human Vaccine Immunotherapy* 16, 9(3).

Greco, R., Michel, M., Guetard, D., Cervantes-Gonzalez, M., Pelucchi, N., Wain-Hobson, S., Sala, F. and Sala, M. (2007) Production of recombinant HIV-1/HBV virus-like particles in *Nicotiana tabacum* and *Arabidopsis thaliana* plants for a bivalent plant-based vaccine. *Vaccine* 25(49), 8228–8240.

Guan, Z.J., Guo, B., Huo, Y.L., Guan, Z.P. and Wei, Y.H. (2010) Overview of expression of hepatitis B surface antigen in transgenic plants. *Vaccine* 28(46), 7351–7362.

Guetard, D., Greco, R., Cervantes Gonzalez, M., Celli, S., Kostrzak, A., Langlade-Demoyen, P., Sala, F., Wain-Hobson, S. and Sala, M. (2008) Immunogenicity and tolerance following HIV-1/HBV plant-based oral vaccine administration. *Vaccine* 26(35), 4477–4485.

Hefferon, K. (2013) Plant-derived pharmaceuticals for the developing world. *Biotechnology Journal* 8(10),1193–202.

Lou, X.-M., Yao, Q.-H., Zhang, Z., Peng, R.-H., Xiong, A.-S. and Wang, H.-K. (2007) Expression of the human hepatitis B virus large surface antigen gene in transgenic tomato plants. *Clinical and Vaccine Immunology* 14(4), 464–469.

Marcondes, J. and Hansen, E. (2008) Transgenic lettuce seedlings carrying hepatitis B virus antigen HBsAg. *Brazilian Journal of Infectious Disease* 12(6), 469–471.

McCormick, A.A., Reddy, S., Reinl, S.J., Cameron, T.I., Czerwinski, D.K., Vojdani, F., Hanley, K.M., Garger, S.J., White, E.L., Novak, J., Barrett, J., Holtz, R.B., Tusé, D. and Levy, R. (2008) Plant-produced idiotype vaccines for the treatment of non-Hodgkin's lymphoma: safety and immunogenicity in a phase I clinical study. *Proceedings of the National Academy of Sciences USA* 105(29), 10131–10136.

Paul, M. and Ma, J.K. (2010) Plant-made immunogens and effective delivery strategies. *Expert Review of Vaccines* 9, 821–833.

Penney, C.A., Thomas, D.R., Deen, S.S. and Walmsley, A.M. (2011) Plant-made vaccines in support of the Millennium Development Goals. *Plant Cell Reports* 30(5), 789–798.

Pniewski, T. (2013) The twenty-year story of a plant-based vaccine against hepatitis B: stagnation or promising prospects? *International Journal of Molecular Sciences* 14, 1978–1998.

Pniewski, T., Kapusta, J., Bociąg, P., Wojciechowicz, J., Kostrzak, A., Gdula, M., Fedorowicz-Strońska, O., Wójcik, P., Otta, H., Samardakiewicz, S., Wolko, B. and Płucienniczak, A. (2011) Low-dose oral immunization with lyophilized tissue of herbicide-resistant lettuce expressing hepatitis B surface antigen for prototype plant-derived vaccine tablet formulation. *Journal of Applied Genetics* 52(2), 125–136.

Pniewski, T., Kapusta, J., Bociąg, P., Kostrzak, A., Fedorowicz-Strońska, O., Czyż, M., Gdula, M., Krajewski, P., Wolko, B. and Płucienniczak, A. (2012) Plant expression, lyophilisation and storage of HBV medium and large surface antigens for a prototype oral vaccine formulation. *Plant Cell Report* 31(3), 585–595.

Qian, B., Shen, H., Liang, W., Guo, X., Zhang, C., Wang, Y., Li, G., Wu, A., Cao, K. and Zhang, D. (2008) Immunogenicity of recombinant hepatitis B virus surface antigen fused with preS1 epitopes expressed in rice seeds. *Transgenic Research* 17(4), 621–631.

Richter, L.J., Thanavala, Y., Arntzen, C.J. and Mason, H.S. (2000) Production of hepatitis surface antigen in transgenic plants for oral immunization. *Nature Biotechnology* 18, 1167–1171.

Rigano, M.M., Mannab, C., Giulin, A., Vitale, A. and Cardi, T. (2009) Plants as biofactories for the production of subunit vaccines against bio-security-related bacteria and viruses. *Vaccine* 27, 3463–3466.

Shchelkunov, S.N. and Shchelkunova, G.A. (2010) Plant-based vaccines against human hepatitis B virus. *Expert Review of Vaccines* 9(8), 947–955.

Streatfield, S.J. (2006) Mucosal immunization using recombinant plant-based oral vaccines. *Methods* 38(2), 150–157.

United Nations (2010) *The Millennium Development Goals Report*. United Nations Department of Economic and Social Affairs (DESA), New York.

Youm, J.W., Won, Y.S., Jeon, J.H., Moon, K.B., Kim, H.C., Shin, K.S., Joung, H. and Kim, H.S. (2010) Antibody responses in mice stimulated by various doses of the potato-derived major surface antigen of hepatitis B virus. *Clinical Vaccine Immunology* 17(12), 2029–2032.

2 Protein Body-inducing Fusions for Recombinant Protein Production in Plants

Sonia Gutiérrez and Rima Menassa*

Agriculture and Agri-Food Canada, London, Ontario, Canada

2.1 Introduction

Plants are considered as a safe, efficient and inexpensive alternative system to produce a wide variety of recombinant proteins such as industrial enzymes, vaccines, antibodies and other biopharmaceuticals (Ma *et al.*, 2005; Knäblein, 2006). This area of plant biotechnology is defined as molecular farming.

Plants possess important advantages over conventional expression systems such as bacteria, yeast, mammalian and insect cells. Their major benefit lies in the low cost of large-scale production, in part because the agricultural systems needed for cultivation are already available, reducing the capital and operating costs (Lico *et al.*, 2005). As the protein synthesis is conserved between animals and plants, recombinant proteins can undergo proper post-translational modifications, folding and assembly. Furthermore, direct oral administration of unprocessed or partially processed plant material is possible due to the absence of human pathogens and endotoxins (Twyman *et al.*, 2003).

The use of plants as bioreactors for the production of recombinant proteins has recently achieved its first success (Maxmen, 2012). The plant-produced taliglucerase alfa for treatment of Gaucher disease was approved in May 2012 by the FDA. This is the first example of an approval for a plant-made therapeutic protein and the first time a major pharmaceutical company embraced this technology.

2.2 *Nicotiana tabacum* as a Plant-based Production Platform

Although a wide range of host plants have been used for molecular farming, tobacco has taken the lead for recombinant protein production for several reasons, including the ease of genetic transformation, high biomass yield (more than 100,000 kg tissue ha^{-1}), and the availability of *Nicotiana* germplasm with diverse agronomic properties and ability to accumulate

* Corresponding author, E-mail: Rima.Menassa@agr.gc.ca

recombinant proteins (Conley *et al.*, 2011b). For example, cultivar I64 has been classified as a high biomass cultivar suitable for production of industrial enzymes (Kolotilin *et al.*, 2013), whereas cv. 81V9 is a low alkaloid cultivar that has been used for direct oral administration of the protein products (Menassa *et al.*, 2007).

Furthermore, biosafety and biocontainment are important considerations in the production of therapeutic proteins. With tobacco, a non-food, non-feed crop, the risk of biologically active therapeutic proteins entering the human and animal food supply is eliminated (Menassa *et al.*, 2001; Twyman *et al.*, 2003). As well, the tobacco expression platform is based on leaves, which minimizes gene leakage into the environment through pollen or seed dispersal (Rymerson *et al.*, 2002).

However, some disadvantages are associated with a leaf-based system, the most important of which are low accumulation levels of some recombinant proteins (usually due to proteolytic degradation or poor stability), and the lack of efficient and scalable protein purification methods from complex leaf protein extracts (Twyman *et al.*, 2003; Conley *et al.*, 2011a,b). Because of low protein expression levels in plants, yield is the biggest hurdle for successful production of recombinant proteins, and it is crucial to address this issue for the commercial success of a tobacco-based expression system.

2.3 Plant Seeds: Organs Naturally Suited for Protein Storage

Cereal seeds are currently well-developed platforms used for the production of recombinant proteins. The main reasons for their success are their high grain yield, their relatively high seed protein content and the existence of well-developed molecular tools for their genetic manipulation (Stoger *et al.*, 2005; Lau and Sun, 2009). Therefore, studying the accumulation of seed storage proteins (SSPs) is expected to provide information that can be used to increase the amount of recombinant proteins in seeds (Kawakatsu and Takaiwa, 2010).

Based on their solubility, plant SSPs have been classified into albumins (water soluble), globulins (salt soluble), prolamins (aqueous alcohol soluble) and glutelins (dilute acid or alkali soluble). Albumins and globulins are the major SSPs of dicots, whereas prolamins and glutelins are the major SSPs in monocots (Stoger *et al.*, 2001; Shewry and Halford, 2002; Kawakatsu and Takaiwa, 2010).

SSPs are synthesized on rough endoplasmic reticulum (rER) membranes and after co-translational cleavage of their N-terminal signal peptide, they are translocated to the ER lumen. Most SSPs are deposited into protein storage vacuoles (PSVs) (Shewry and Halford, 2002). Alternatively, prolamins accumulate within the ER lumen and form ER-derived accretions called protein bodies (PB), which are surrounded by a membrane of ER origin (Larkins and Hurkman, 1978; Coleman *et al.*, 1996; Herman and Larkins, 1999; Kawakatsu and Takaiwa, 2010). Prolamins lack the classical C-terminal KDEL or HDEL motif for retention within ER. Therefore, it appears they have intrinsic structural features responsible for their retention in the ER (Munro and Pelham, 1987; Coleman *et al.*, 1996; Herman and Larkins, 1999). The ability to sequester proteins in these large and dense PBs seems to be the key to enhance accumulation of proteins in seeds (Galili, 2004; Torrent *et al.*, 2009a). Unfortunately, the mechanisms behind transport and deposition of prolamins in PBs are still not clear (Llop-Tous *et al.*, 2010).

Although the use of seeds as platforms for molecular farming was initially popular, concerns by the public and regulatory agencies have led to an effective worldwide decrease in their use. Specifically, regulators are concerned about the spread of transgenes to wild plants and transgenic seeds entering human or animal food chains. To circumvent this, the use of non-food leaf-based production systems for the generation of recombinant proteins has been suggested (Rybicki, 2010). The use of the natural mechanism of protein accumulation in seeds, combined with leaf-based production systems, offers a valuable option for an enhanced production of recombinant proteins.

2.4 Adapting Seed Mechanisms to Improve Leaf-based Production Systems

The mechanisms behind prolamin aggregation, which are responsible for PB assembly, as well as the mechanisms that determine their ER localization are only partially understood (Torrent *et al.*, 2009a; Llop-Tous *et al.*, 2010). Prolamins are characterized by their high content in proline and their solubility in alcohol, which reflects their general hydrophobic nature (Shewry and Halford, 2002). Studies in maize showed that PBs were directly formed in the lumen of the ER and contained four structurally different maize prolamins: α-, β-, δ- and γ-zein (Coleman *et al.*, 1996). Zeins are distributed heterogeneously in PBs; initially the small PBs consist of β- and γ-zein (sulfur-rich prolamins) aggregates (Lending and Larkins, 1989). This is followed by penetration and accumulation of α- and δ-zeins into the centre of the PB, expanding it until it reaches a diameter of 1 to 2 μm. As the centre of the PB is filled, β- and γ-zeins are concentrated toward the periphery of PBs (Lending and Larkins, 1989; Kim *et al.*, 2002). This spatial organization of zeins in PBs suggests that each of these proteins has specific properties, responsible for ER retention and localization within the PBs (Coleman *et al.*, 1996).

Further characterization of γ-zein showed that it facilitates the assembly of the other PB-associated zeins and stabilizes them. Moreover, heterologous expression of γ-zein in leaves promotes the formation of PB-like structures inside the ER, suggesting that structural motifs within γ-zein may be responsible for PB induction. It was later shown that the proline-rich repetitive sequences (PPPVHL)$_8$ present at its N-terminus and a central Pro-X domain are responsible for γ-zein's ability to self-assemble and be retained within the ER (Geli *et al.*, 1994; Torrent *et al.*, 2009a). Therefore, a synthetic peptide of the N-terminal proline-rich repetitive sequence

(PPPVHL)$_8$ of γ-zein was synthesized and tested *in vitro*. Surprisingly, this peptide was able to self-assemble and form cylindrical micelles, which are globular aggregates where the hydrophilic regions of the peptide are exposed to the surface and the hydrophobic regions of the peptide are clumped together in the centre of the aggregate (Kogan *et al.*, 2002; Kim and Xu, 2008).

Consequently, a fusion protein-based system designed to accumulate recombinant proteins in ER-derived PBs in plants was developed (Torrent *et al.*, 2009a). The approach relies on fusion proteins that contain the γ-zein signal peptide, the γ-zein proline-rich domain (Zera®; developed by ERA Biotech, Barcelona, Spain) and the protein of interest fused to the C-terminus of the proline-rich region (Torrent *et al.*, 2009b). The formation of γ-zein-induced PBs was confirmed in plant vegetative tissue and in several non-plant eukaryotic systems (i.e. fungal, insect and mammalian cells), where enhanced recombinant protein accumulation by using Zera was also evidenced (Torrent *et al.*, 2009b; Llop-Tous *et al.*, 2010; StorPro organelles http://www.erabiotech.com). This new system allows the expression of heterologous recombinant proteins in non-seed tissues with three main advantages: (i) proteins remain stable inside membrane-bound PBs; (ii) the induced PBs are dense organelles which can be easily purified by centrifugation on density gradients; and (iii) the presence of these PBs does not affect normal growth and development of the hosts (Torrent *et al.*, 2009a, b).

2.5 Fusion Partners as Enhancers of Recombinant Protein Accumulation and PB Inducers

In recent years, fusion protein technology has been extended to other fusion tags for enhancing recombinant protein accumulation in plants. Among these, proteins fused with elastin-like polypeptides (ELP) and hydrophobin (HFBI) have been very successful when targeted to the ER and have shown increased accumulation and potential in non-chromatographic purification (Joensuu *et al.*, 2010; Conley *et al.*, 2011a).

2.5.1 Elastin-like polypeptides as fusion partners

ELPs are synthetic biopolymers composed of a repeating amino acid sequence 'VPGXG' where X can be any amino acid except proline. These repeats occur naturally in all mammalian elastin proteins (Raju and Anwar, 1987). In aqueous solution, ELPs undergo a reversible inverse phase transition from soluble protein to insoluble hydrophobic aggregates that form β-spiral structures when heated above their transition temperature (T_t) (Urry, 1988). This thermally responsive property of ELPs is also transferred to fusion partners, providing a scalable non-chromatographic method for protein purification called 'Inverse Transition Cycling' (ITC) (Meyer and Chilkoti, 1999). It was previously shown by ELP expression in bacteria that the T_t of an ELP tag varies with its sequence, length and concentration. Because a significant increase in T_t was observed with short ELP tags, ELPs containing 90–180 repeats have been used for ITC purification studies (Meyer and Chilkoti, 1999).

In plants, synthetic ELP tags have been successfully used in fusion with other proteins for high level production and purification by ITC. Some examples include fusions with cytokines (Lin *et al.*, 2006; Patel *et al.*, 2007; Conley *et al.*, 2009a), antibodies (Floss *et al.*, 2008; Conley *et al.*, 2009a; Joensuu *et al.*, 2009) and spider silk (Scheller *et al.*, 2004; Patel *et al.*, 2007). All of these ELP fusions were tested in stable transgenic tobacco plants and significantly enhanced accumulation of the recombinant proteins. However, the concentration of recombinant protein was inversely proportional to the size of ELP tag both in plants and in bacteria (Meyer and Chilkoti, 2004; Conley *et al.*, 2009a). In plants, 30 pentapeptide repeats appear to be the best compromise between the positive effects of smaller ELP tags on accumulation, and the ability of larger ELP tags to aggregate during ITC. As well, positioning the ELP tag at the C-terminal of the fusion enhances accumulation, while N-terminal ELP does not (Christensen *et al.*, 2009; Conley *et al.*, 2009a).

To understand better the role of ELP tags in the enhanced accumulation of fusion proteins in tobacco plants, an ER-targeted green fluorescent protein (GFP)–ELP fusion was used in transient expression assays and uncovered the presence of novel vesicles in leaves of *Nicotiana benthamiana*, resembling cereal seed protein bodies (Fig. 2.1A, B; Conley *et al.*, 2009b). Electron microscopy (EM) revealed that these vesicles are surrounded by a membrane studded with ribosomes, and co-localization of an ER-resident molecular chaperone binding protein (BiP) and an ER-targeted yellow fluorescent protein (YFP)-ELP fusion confirmed the ER origin of these vesicles, which were identified as protein bodies (Fig. 2.1C, D) (Conley *et al.*, 2009b).

The current experimental evidence indicates that ELP tags increase the stability and solubility of target proteins, leading to an increase in the accumulation levels of the recombinant proteins when they are retained in the ER. A high localized accumulation of the recombinant protein in the ER lumen may then induce formation of PBs, thus protecting the recombinant protein from degradation. This effect seems to be explained by specific biochemical properties of ELP, reminiscent of seed prolamins like γ-zein. Indeed, ELPs and γ-zeins are hydrophobic and proline-rich, characteristics responsible for the self-assembly of prolamins in a wide range of hosts (Llop-Tous *et al.*, 2010).

2.5.2 Hydrophobin I as a fusion partner

Hydrophobins are small, secreted proteins present in filamentous fungi such as *Schizophyllum commune* and *Trichoderma reesei* (Nakari-Setala *et al.*, 1996). These proteins are involved in fungal growth, development, and adaptation to the external environment. They contribute to surface

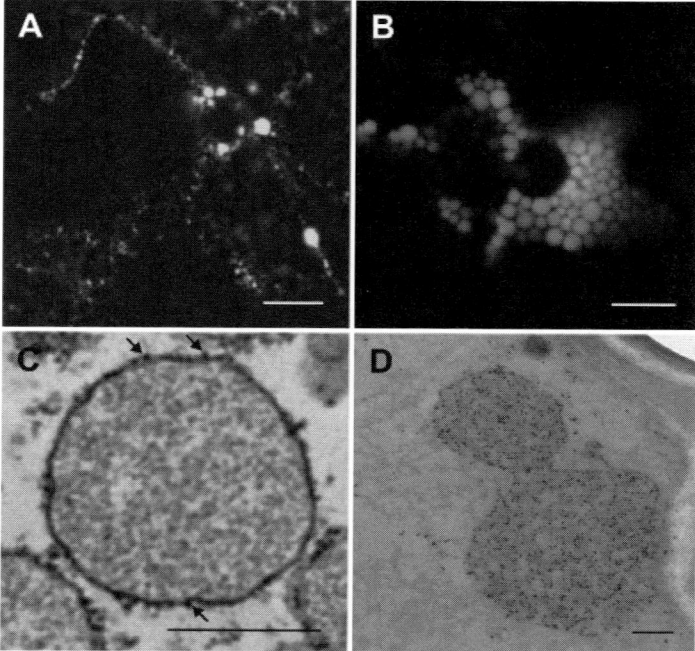

Fig. 2.1. Protein bodies induced in tobacco leaves by the expression of an ER-targeted GFP–ELP fusion. (A) PBs were closely associated with the ER network as small punctuate structures early on in the PB formation process. (B) Different sizes of PBs. The PBs cluster together within the cell, although the distribution pattern is variable. (C) PB surrounded by a membrane decorated with ribosomes (indicated with arrows). The presence of the ribosomes on the membrane suggests a rough ER origin. (D) Immunogold localization confirmed the presence of GFP–ELP fusion protein inside the PBs using anti-ELP antibodies. Confocal microscopy images courtesy of Reza Saberianfar. Electron microscopy images courtesy of Dr Jussi Joensuu. Bars = 10 μm (A and B), 500 nm (C and D).

hydrophobicity, and thus help the dispersal and survival of fungal spores and adhesion to host surfaces (Hakanpaa *et al.*, 2004; Linder *et al.*, 2004). Their main feature is the presence of eight cysteine residues, which form four intramolecular disulfide bonds and fold into a stable tertiary structure with an exposed 'hydrophobic patch' on one side of the protein. This is an uncommon characteristic as hydrophobic residues are usually buried in the core of hydrophobic proteins, thus stabilizing their conformation (Linder, 2009). Thus hydrophobins are amphipathic molecules, with a hydrophilic and a hydrophobic region, similar to the structure of surfactants (Linder, 2009; Joensuu *et al.*, 2010; Conley *et al.*, 2011a).

Hydrophobins have been divided into two classes (class I and class II) based on the presence of hydrophilic and hydrophobic amino acid residues in their protein sequence. Members of class I are characterized by their highly insoluble aggregates in aqueous solution, whereas aggregates of the members of class II are easier to dissolve (Linder, 2009). Due to their structural properties, hydrophobins can self-assemble into an amphipathic protein membrane at hydrophilic-hydrophobic interfaces, and they have a very high affinity for surfactants (Wösten and de Vocht, 2000; Wang *et al.*, 2005).

Hydrophobin I (HFBI) from *Trichoderma reesei* is a class II hydrophobin (Nakari-Setala *et al.*, 1996), and can be purified by a two-step surfactant-based aqueous two-phase system (ATPS) (Linder *et al.*, 2001). During ATPS, a surfactant is added to the crude protein extract, which concentrates HFBI inside micelles and partitions them towards the surfactant phase (Lahtinen *et al.*, 2008). HFBI can then be recovered from the surfactant phase with a non-denaturing organic solvent, such as isobutanol (Linder *et al.*, 2001). When fused to other proteins, HFBI can alter the hydrophobicity of the fusion partner allowing for simple, scalable and inexpensive purification using ATPS (Linder *et al.*, 2004). HFBI has been used to over-express and purify recombinant proteins from *Trichoderma* sp., insect cells and plant tissues (Linder *et al.*, 2001; Lahtinen *et al.*, 2008; Joensuu *et al.*, 2010). Recovery of an ER targeted GFP–HFBI fusion protein from plant leaf extracts showed that it selectively recovered up to 91% of the fusion protein (Joensuu *et al.*, 2010).

In plants, the HFBI tag was shown to significantly increase the accumulation of GFP and to induce formation of PBs comparable to the ones observed with the ELP tag (Fig. 2.2). Transmission electron microscopy revealed that PBs are surrounded by a membrane studded with ribosomes. Immunogold labelling for GFP revealed specific gold decoration of the PBs.

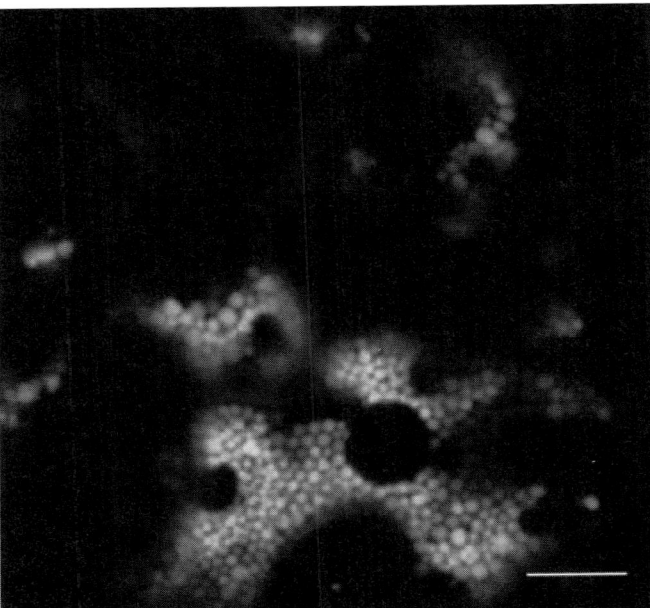

Fig. 2.2. Expression of an ER-targeted GFP–HFBI fusion induces the formation of protein bodies in tobacco leaves. Confocal microscopy image of GFP–HFBI transient expression in *N. benthamiana*. Leaf epidermal cell accumulating GFP-HFBI in PBs. Bar = 10 μm. Confocal microscopy image courtesy of Reza Saberianfar.

Interestingly, a protective effect of the HFBI tag in the infiltrated tissue was observed. Although over-expression in transient systems usually leads to necrotic lesions 4 days post-infiltration (dpi), upon infiltration of GFP–HFBI the tissue remained healthy for up to 10 dpi, and recombinant protein accumulation remained stable. A possible explanation for this protective effect may be the sequestration of the fusion proteins in PBs. It is thought that packaging the recombinant protein into PBs can prevent possible negative effects of over-expression of a foreign protein in the plant cell (Joensuu *et al.*, 2010).

2.6 Elastin-like Polypeptide and HFBI I Fusions in Transgenic Tobacco

The positive effect of ELP and HFBI on accumulation levels of their fusion partners and the finding that the fusions do not, in most cases, affect the biological activity of their fusion partners, have triggered a renewed interest in high-level production and purification of recombinant proteins. However, most studies to date have been carried out using transient expression by agro-infiltration, whereas using plants as bioreactors for recombinant proteins would be more efficient if stable transgenic plants are generated. This would facilitate scaling-up for high volume production.

In an effort to assess this technology in a constitutive environment, we evaluated the effect of the ELP and HFBI fusion tags for recombinant protein accumulation and PB formation in stable transgenic *N. tabacum* plants. By transforming ER-targeted GFP, GFP–HFBI and GFP–ELP fusions in two tobacco cultivars, we have shown that these tags increase the accumulation of GFP and induce the formation of PBs in leaves of stable transgenic plants in both cultivars. Furthermore, we found that these tags induce the formation of PBs in a concentration-dependent manner, where a minimum level of recombinant protein accumulation is required for PBs to appear (Gutierrez *et al.*, 2013). Additionally, we demonstrated that post-transcriptional gene silencing (PTGS) can play a role in expression and accumulation of recombinant proteins in transgenic plants, since agro-infiltrating low-accumulating GFP–HFBI transgenic lines with p19, a suppressor of PTGS (Fagard and Vaucheret, 2000; Silhavy *et al.*, 2002; Brodersen and Voinnet, 2006), increased accumulation levels of GFP–HFBI (Gutierrez *et al.*, 2013).

2.7 Conclusions

The use of PB-inducing fusion tags in transient expression and in stable transgenic plants is feasible and promising. Both of these tags increase the accumulation levels of the recombinant protein and induce the formation of PBs, while providing a fast and efficient purification method.

Acknowledgement

Work in the authors' lab was supported by A-base grants from Agriculture and Agri-Food Canada.

References

Brodersen, P. and Voinnet, O. (2006) The diversity of RNA silencing pathways in plants. *Trends in Genetics* 22, 268–280.

Christensen, T., Amiram, M., Dagher, S., Trabbic-Carlson, K., Shamji, M.F., Setton, L.A. and Chilkoti, A. (2009) Fusion order controls expression level and activity of elastin-like polypeptide fusion proteins. *Protein Science: a Publication of the Protein Society* 18, 1377–1387.

Coleman, C.E., Herman, E.M., Takasaki, K. and Larkins, B.A. (1996) The maize γ-zein sequesters α-zein and stabilizes its accumulation in protein bodies of transgenic tobacco endosperm. *The Plant Cell* 8, 2335–2345.

Conley, A.J., Joensuu, J.J., Jevnikar, A.M., Menassa, R. and Brandle, J.E. (2009a) Optimization of elastin-like polypeptide fusions for expression and purification of recombinant proteins in plants. *Biotechnology and Bioengineering* 103, 562–573.

Conley, A.J., Joensuu, J.J., Menassa, R. and Brandle, J.E. (2009b) Induction of protein body formation in plant leaves by elastin-like polypeptide fusions. *BMC Biology* 7, 48.

Conley, A.J., Joensuu, J.J., Richman, A. and Menassa, R. (2011a) Protein body-inducing fusions for high-level production and purification of recombinant proteins in plants. *Plant Biotechnology Journal* 9, 419–433.

Conley, A.J., Zhu, H., Le, L.C., Jevnikar, A.M., Lee, B.H., Brandle, J.E. and Menassa, R. (2011b) Recombinant protein production in a variety of *Nicotiana* hosts: a comparative analysis. *Plant Biotechnology Journal* 9, 434–444.

Fagard, M. and Vaucheret, H. (2000) (Trans)gene silencing in plants: how many mechanisms? *Annual Review of Plant Physiology and Plant Molecular Biology* 51, 167–194.

Floss, D.M., Sack, M., Stadlmann, J., Rademacher, T., Scheller, J., Stoger, E., Fischer, R. and Conrad, U. (2008) Biochemical and functional characterization of anti-HIV antibody-ELP fusion proteins from transgenic plants. *Plant Biotechnology Journal* 6, 379–391.

Galili, G. (2004) ER-derived compartments are formed by highly regulated processes and have special functions in plants. *Plant Physiology* 136, 3411–3413.

Geli, M.I., Torrent, M. and Ludevid, D. (1994) Two structural domains mediate two sequential events in gamma-zein targeting: Protein endoplasmic reticulum retention and protein body formation. *The Plant Cell* 6, 1911–1922.

Gutierrez, S.P., Saberianfar, R., Kohalmi, S.E. and Menassa, R. (2013) Protein body formation in stable transgenic tobacco expressing elastin-like polypeptide and hydrophobin fusion proteins. *BMC Biotechnology* 13, 40.

Hakanpaa, J., Paananen, A., Askolin, S., Nakari-Setala, T., Parkkinen, T., Penttila, M., Linder, M.B. and Rouvinen, J. (2004) Atomic resolution structure of the HFBII hydrophobin, a self-assembling amphiphile. *The Journal of Biological Chemistry* 279, 534–539.

Herman, E.M. and Larkins, B.A. (1999) Protein storage bodies and vacuoles. *The Plant Cell* 11, 601–613.

Joensuu, J.J., Brown, K.D., Conley, A.J., Clavijo, A., Menassa, R. and Brandle, J.E. (2009) Expression and purification of an anti-Foot-and-mouth disease virus single chain variable antibody fragment in tobacco plants. *Transgenic Research* 18, 685–696.

Joensuu, J.J., Conley, A.J., Lienemann, M., Brandle, J.E., Linder, M.B. and Menassa, R. (2010) Hydrophobin fusions for high-level transient protein expression and purification in *Nicotiana benthamiana*. *Plant Physiology* 152, 622–633.

Kawakatsu, T. and Takaiwa, F. (2010) Cereal seed storage protein synthesis: fundamental processes for recombinant protein production in cereal grains. *Plant Biotechnology Journal* 8, 939–953.

Kim, C.S., Woo Y.M., Clore, A.M., Burnett, R.J., Carneiro, N.P. and Larkins, B.A. (2002) Zein protein interactions, rather than the asymmetric distribution of zein mRNAs on endoplasmic reticulum membranes, influence protein body formation in maize endosperm. *Plant Cell* 14, 655–672.

Kim, S. and Xu, J. (2008) Aggregate formation of zein and its structural inversion in aqueous ethanol. *Journal of Cereal Science* 47, 1–5.

Knäblein, J. (2006) Plant-based expression of biopharmaceuticals. In: *Encyclopedia of Molecular Cell Biology and Molecular Medicine.* Wiley-VCH Verlag GmbH & Co. KGaA, doi:10.1002/3527600906. mcb.200400120.

Kogan, M.J., Dalcol, L., Gorostiza, P., Lopez-Iglesias, C., Pons, R., Pons, M., Sanz, F. and Giralt, E. (2002) Supramolecular properties of the proline-rich γ-zein N-terminal domain. *Biophysical Journal* 83, 1194–1204.

Kolotilin, I., Kaldis, A., Pereira, E.O., Laberge, S. and Menassa, R. (2013) Optimization of transplastomic production of hemicellulases in tobacco: effects of expression cassette configuration and tobacco cultivar used as production platform on recombinant protein yields. *Biotechnology for Biofuels* 6, 65.

Lahtinen, T., Linder, M.B., Nakari-Setala, T. and Oker-Blom, C. (2008) Hydrophobin (HFBI): A potential fusion partner for one-step purification of recombinant proteins from insect cells. *Protein Expression and Purification* 59, 18–24.

Larkins, B.A. and Hurkman, W.J. (1978) Synthesis and deposition of zein in protein bodies of maize endosperm. *Plant Physiology* 62, 256–263.

Lau, O.S. and Sun, S.S. (2009) Plant seeds as bioreactors for recombinant protein production. *Biotechnology Advances* 27, 1015–1022.

Lending, C.R. and Larkins, B.A. (1989) Changes in the zein composition of protein bodies during maize endosperm development. *The Plant Cell* 1, 1011–1023.

Lico, C., Desiderio, A., Banchieri, S. and Benvenuto, E. (2005) Plants as biofactories: production of pharmaceutical recombinant proteins. In: Tuberosa, R., Phillips, R.L. and Gale, M. (eds) *In the Wake of the Double Helix: from the Green Revolution to the Gene Revolution: Proceedings of an International Congress, Bologna, Italy, May 27–31, 2003.* Avenue media, Bologna, Italy, pp.577–593.

Lin, M., Rose-John, S., Grotzinger, J., Conrad, U. and Scheller, J. (2006) Functional expression of a biologically active fragment of soluble gp130 as an ELP-fusion protein in transgenic plants: purification via inverse transition cycling. *Biochemical Journal* 398, 577–583.

Linder, M., Selber, K., Nakari-Setala, T., Qiao, M., Kula, M.R. and Penttila, M. (2001) The hydrophobins HFBI and HFBII from *Trichoderma reesei* showing efficient interactions with nonionic surfactants in aqueous two-phase systems. *Biomacromolecules* 2, 511–517.

Linder, M.B. (2009) Hydrophobins: Proteins that self assemble at interfaces. *Current Opinion in Colloid & Interface Science* 14, 356–363.

Linder, M.B., Qiao, M., Laumen, F., Selber, K., Hyytia, T., Nakari-Setala, T. and Penttila, M.E. (2004) Efficient purification of recombinant proteins using hydrophobins as tags in surfactant-based two-phase systems. *Biochemistry* 43, 11873–11882.

Llop-Tous, I., Madurga, S., Giralt, E., Marzabal, P., Torrent, M. and Ludevid, M.D. (2010) Relevant elements of a Maize γ-zein domain involved in protein body biogenesis. *The Journal of Biological Chemistry* 285, 35633–35644.

Ma, J.K., Barros, E., Bock, R., Christou, P., Dale, P.J., Dix, P.J., Fischer, R., Irwin, J., Mahoney, R., Pezzotti, M., Schillberg, S., Sparrow, P., Stoger, E. and Twyman, R.M. (2005) Molecular farming for new drugs and vaccines. Current perspectives on the production of pharmaceuticals in transgenic plants. *EMBO Reports* 6, 593–599.

Maxmen, A. (2012) Drug-making plant blooms. *Nature* 485, 160.

Menassa, R., Nguyen, V., Jevnikar, A. and Brandle, J. (2001) A self-contained system for the field production of plant recombinant interleukin-10. *Molecular Breeding* 8, 177–185.

Menassa, R., Du, C., Yin, Z.Q., Ma, S., Poussier, P., Brandle, J. and Jevnikar, A.M. (2007) Therapeutic effectiveness of orally administered transgenic low-alkaloid tobacco expressing human interleukin-10 in a mouse model of colitis. *Plant Biotechnology Journal* 5, 50–59.

Meyer, D.E. and Chilkoti, A. (1999) Purification of recombinant proteins by fusion with thermally-responsive polypeptides. *Nature Biotechnology* 17, 1112–1115.

Meyer, D.E. and Chilkoti, A. (2004) Quantification of the effects of chain length and concentration on the thermal behavior of elastin-like polypeptides. *Biomacromolecules* 5, 846–851.

Munro, S. and Pelham, H.R. (1987) A C-terminal signal prevents secretion of luminal ER proteins. *Cell* 48, 899–907.

Nakari-Setala, T., Aro, N., Kalkkinen, N., Alatalo, E. and Penttila, M. (1996) Genetic and biochemical characterization of the *Trichoderma reesei* hydrophobin HFBI. *European Journal of Biochemistry* 235, 248–255.

Patel, J., Zhu, H., Menassa, R., Gyenis, L., Richman, A. and Brandle, J. (2007) Elastin-like polypeptide fusions enhance the accumulation of recombinant proteins in tobacco leaves. *Transgenic Research* 16, 239–249.

Raju, K. and Anwar, R.A. (1987) Primary structures of bovine elastin a, b, and c deduced from the sequences of cDNA clones. *The Journal of Biological Chemistry* 262, 5755–5762.

Rybicki, E.P. (2010) Plant-made vaccines for humans and animals. *Plant Biotechnology Journal* 8, 620–637.

Rymerson, R., Menassa, R. and Brandle, J. (2002) Tobacco, a platform for the production of recombinant proteins. In: Erickson, L., Yu, W.-J., Brandle, J. and Rymerson, R. (eds) *Molecular Farming of Plants and Animals for Human and Veterinary Medicine*. Kluwer Academic Publishers, Dorcrecht, the Netherlands, pp. 1–32.

Scheller, J., Henggeler, D., Viviani, A. and Conrad, U. (2004) Purification of spider silk-elastin from transgenic plants and application for human chondrocyte proliferation. *Transgenic Research* 13, 51–57.

Shewry, P.R. and Halford, N.G. (2002) Cereal seed storage proteins: structures, properties and role in grain utilization. *Journal of Experimental Botany* 53, 947–958.

Silhavy, D., Molnar, A., Lucioli, A., Szittya, G., Hornyik, C., Tavazza, M. and Burgyan, J. (2002) A viral protein suppresses RNA silencing and binds silencing-generated, 21- to 25-nucleotide double-stranded RNAs. *EMBO Journal* 21, 3070–3080.

Stoger, E., Parker, M., Christou, P. and Casey, R. (2001) Pea legumin overexpressed in wheat endosperm assembles into an ordered paracrystalline matrix. *Plant Physiology* 125, 1732–1742.

Stoger, E., Ma, J.K., Fischer, R. and Christou, P. (2005) Sowing the seeds of success: pharmaceutical proteins from plants. *Current Opinion in Biotechnology* 16, 167–173.

Torrent, M., Llompart, B., Lasserre-Ramassamy, S., Llop-Tous, I., Bastida, M., Marzabal, P., Westerholm-Parvinen, A., Saloheimo, M., Heifetz, P.B. and Ludevid, M.D. (2009a) Eukaryotic protein production in designed storage organelles. *BMC Biology* 7, 5.

Torrent, M., Llop-Tous, I. and Ludevid, M.D. (2009b) Protein body induction: a new tool to produce and recover recombinant proteins in plants. *Methods in Molecular Biology (Clifton, New Jersey)* 483, 193–208.

Twyman, R.M., Stoger, E., Schillberg, S., Christou, P. and Fischer, R. (2003) Molecular farming in plants: host systems and expression technology. *Trends in Biotechnology* 21, 570–578.

Urry, D.W. (1988) Entropic elastic processes in protein mechanisms. I. Elastic structure due to an inverse temperature transition and elasticity due to internal chain dynamics. *Journal of Protein Chemistry* 7, 1–34.

Wang, X., Shi, F., Wosten, H.A., Hektor, H., Poolman, B. and Robillard, G.T. (2005) The SC3 hydrophobin self-assembles into a membrane with distinct mass transfer properties. *Biophysical Journal* 88, 3434–3443.

Wösten, H.A. and de Vocht, M.L. (2000) Hydrophobins, the fungal coat unravelled. *Biochimica et Biophysica Acta* 1469, 79–86.

3 Expression of Recombinant Proteins in Plant Cell Culture

Somen Nandi[1,2]* and Karen A. McDonald[1,3]

[1]Global HealthShare Initiative; [2]Department of Molecular and Cellular Biology; [3]Department of Chemical Engineering & Materials Science, University of California, Davis, California

3.1 Introduction

Plants possess an exceptional biosynthetic capacity for protein accumulation and production. This can be achieved by using very simple chemically defined media that support biomass production and ultimately production of the target molecule. The potential for cost-effective production of bioactive recombinant proteins is well documented (Sharma and Sharma, 2009; Obembe et al., 2011; Paul and Ma, 2011; Xu et al., 2012). In most cases, plant cells carry out post-translational modifications that are vital for most complex eukaryotic proteins. This provides flexibility in bioproduction platforms that address production scale, cost, safety and regulatory issues. Plant cells generally cannot harbour human and animal pathogens, which brings significant advantage in terms of increased safety for target molecules as well as reduced problems in manufacturing compared to mammalian cell-based production systems (Pogue et al., 2010; Xu et al., 2011). These biosafety advantages also impact commercial aspects; they reduce purification costs and minimize risks associated with potential production shut-downs, facility decontamination and supply limitations leading to unmet patient/customer demand (Wilken and Nikolov, 2012). The costs for purification of any protein, either from plant cells or from microbial or mammalian systems, will be comparable, but in general plant systems require much less initial capital investment to initiate commercial level production (Wilken and Nikolov, 2012). In contrast to other expression systems such as bacterial, mammalian cell and yeast, plant expression systems encompass diverse forms including whole-plants, suspension cells, hairy roots, moss, duckweed, microalgae, etc. (Xu et al., 2012). Each of the platforms has its own strengths and weaknesses (see Fig. 2 in Xu et al., 2012) and is often best suited for certain classes of recombinant molecules based on the structure and character of the target itself as well as the market, scale, cost, and upstream and downstream processing constraints of the particular product. An array of plant species and different tissues can serve as hosts for plant-based bioproduction that comprise platforms

* Corresponding author, E-mail: snandi@ucdavis.edu

ranging from *in vitro* cell and plant tissue cultures to whole plants grown in greenhouses, indoor growth facilities or in the field. Availability and research advancement of such diverse plant-based expression systems makes it possible to respond rapidly to produce novel recombinant proteins that can be customized and meet the need, and provides an opportunity for expression of recombinant proteins in plant cells. This review highlights the advantages and challenges associated with each type of plant cell culture production system strategy. We have also emphasized the important issues related to regulation and progress toward product commercialization in the developed and developing world.

3.2 Different Plant Cell Culture Platforms to Produce Recombinant Molecules

Like microorganisms, undifferentiated clusters of plant cells (calli) can be dispersed and propagated in a liquid medium to generate stable cell suspension cultures, while retaining the same production capacity as a whole plant. *In vitro* culture systems are characterized by the fact that plant biomass is cultured in confined bioreactors under sterile conditions for the large-scale production of recombinant proteins. In most cases, the plant cells grow heterotrophically, using sugar as a carbon and energy source and consuming oxygen for respiration, although some systems described below utilize photobioreactors in which light and/or carbon dioxide can be growth rate limiting factors. But at present, plant cell cultures need not be confined in sterile conditions per se, at least during the downstream production cycles. There are multiple options such as cell-suspension culture, hairy roots, moss and microalgae that can be utilized to express and produce an array of foreign heterologous molecules. Following is a brief overview of those options and advantages and limitations of the various systems.

3.2.1 Plant cell suspension culture

Historically, human serum albumin was the first complex recombinant protein that was successfully expressed in tobacco cell culture (Sijmons *et al.*, 1990). After that, an array of bioactive molecules including antibodies, vaccines, hormones, growth factors and cytokines have been produced in tobacco (*Nicotiana tabacum* or NT) plant cells (Matsumoto *et al.*, 1993; Magnuson *et al.*, 1998; Smith *et al.*, 2002; Kwon *et al.*, 2003b; Lienard *et al.*, 2007). Tobacco BY-2 (Bright Yellow-2) and NT-1 (*Nicotiana tabacum*-1) cells are the most frequently chosen host lines used for recombinant protein expression (Xu *et al.*, 2011). These closely related cell lines are fast-growing cell culture systems where growth rate can reach up to 0.044 h^{-1} corresponding to a doubling time of about 16 h (Hellwig *et al.*, 2004; Su and Lee, 2007; Ozawa and Takaiwa, 2010; Tremblay *et al.*, 2010).

In suspension culture, plant cells typically grow as aggregates (Fig. 3.1). As they divide, the cell wall keeps the daughter cell attached, although

(A)
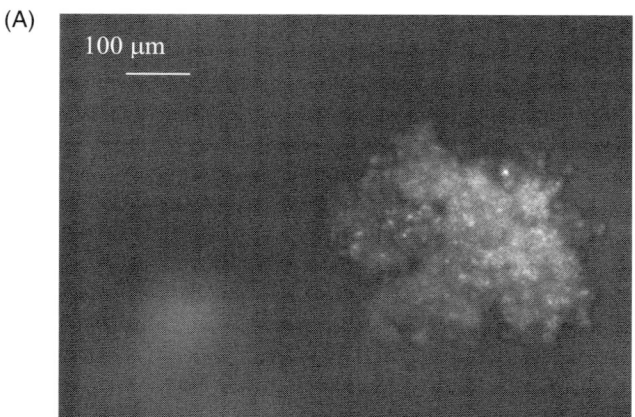

(B)
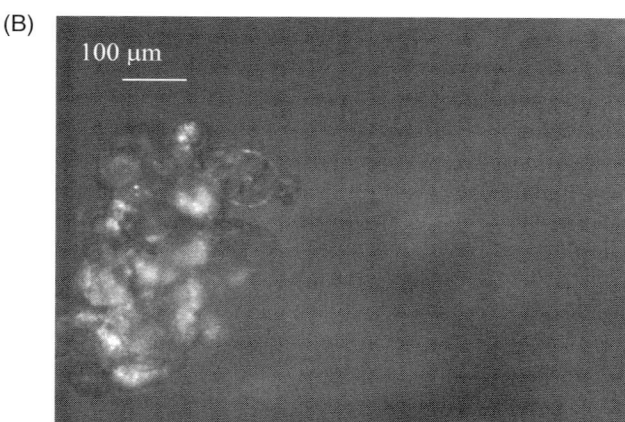

(C)
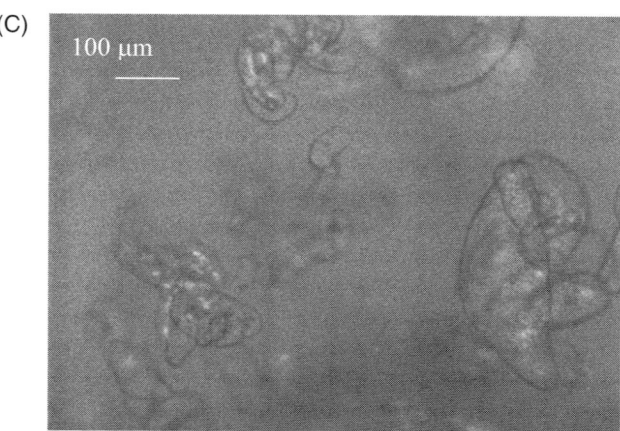

Fig. 3.1. Aggregates in plant cell suspension culture: (A) rice, (B) cucumber and (C) tobacco.

agitation/mixing within the bioreactor can cause aggregates to break apart. Aggregation can sometimes lead to mass transfer limitations wherein cells in the interior of the clump experience a different microenvironment (nutrients, oxygen, etc.), compared to cells on the surface, but the lower metabolic rates of plant cells compared with microbial or fungal systems alleviate this problem. An added benefit of aggregation is that cell/liquid separation can be easily accomplished using gravity sedimentation simply by turning off the agitation and aeration and allowing the aggregates to settle within the bioreactor (Fig. 3.2). This is particularly useful if the product is secreted and stable in the culture media since it allows semicontinuous or perfusion operation, which generally results in higher volumetric productivities.

To date tobacco or *Nicotiana* hosts have the most established history for the production of recombinant proteins in the plant (Conley *et al.*, 2011). An extensive number of proteins have been produced in tobacco cell culture systems over the last decade (Tremblay *et al.*, 2010; Xu *et al.*, 2012). A wide array of biologically active recombinant proteins have been successfully expressed in plant cells, particularly in BY-2 and NT-1. These strains are fast-growing, robust and readily undergo *Agrobacterium*-mediated transformation and cell cycle synchronization (Hellwig *et al.*, 2004; Su and Lee, 2007; Ozawa and Takaiwa, 2010; Tremblay *et al.*, 2010). One bottleneck in exploiting plant cell suspension cultures for commercial purposes has been low productivity. Exceptions are expression of human proteins such as α1-antitrypsin, interleukin-12 and hGM-CSF in rice cells using a sucrose-inducible RAmy3D promoter (Huang *et al.*, 2001; Shin *et al.*, 2003, 2010; McDonald *et al.*, 2005; Trexler *et al.*, 2005), where up to 247 mg l^{-1} was achieved (McDonald *et al.*, 2005). However, the growth rates and characteristics of rice cell lines are inferior to those of tobacco BY-2 and NT-1 cell lines (Hellwig *et al.*, 2004) and the viability of rice cells is significantly decreased when cultivated in a sugar-free medium used for induction of a metabolically regulated promoter (Huang and McDonald, 2009).

Beside tobacco cells, other plant cell culture systems have been successfully utilized for expressing bioactive molecules. Examples include rice (Torres *et al.*, 1999; Huang *et al.*, 2001, 2002, 2005; Shin *et al.*, 2010), soybean (Smith *et al.*, 2002), lucerne (Daniell and Edwards, 1995), tomato (Kwon *et al.*, 2003a) and carrot suspension cell cultures (Shaaltiel *et al.*, 2007), as well as Siberian ginseng (*Acanthopanax senticosus*) (Jo *et al.*, 2006). In some cases plant cell lines derived from common edible crop species might be more favourable than tobacco cells in terms of product levels and regulatory compliance (Hellwig *et al.*, 2004). A recent review by Huang and McDonald (2009) and a comprehensive list of recombinant pharmaceutical proteins expressed using plant cell-based platforms has been presented by Xu *et al.* (Table 1 in Xu *et al.*, 2011).

(A)

(B)

(C)

Fig. 3.2. Plant cell aggregate sedimentation in a 5-l bioreactor. Time after turning off agitation and aeration: (A) t = 0 min, (B) t = 8 min and (C) t = 26 min.

3.2.2 Culture of hairy roots

Hairy roots are generated by infection of plants with *Agrobacterium rhizogenes* (Shanks and Morgan, 1999). The induced root tissues can be grown indefinitely *in vitro* and have become an alternative production platform for proteins and plant metabolites (Giri and Narasu, 2000; Guillon *et al.*, 2006). Hairy roots can be cultured in a controlled environment and it is possible to achieve extracellular secretion from cultured hairy roots, or rhizosecretion. This system also offers a simplified method for the recovery of foreign proteins in a defined medium. The first plant recombinant protein, a full-length murine IgG1, was successfully produced in hairy root culture N10 over a decade ago (Wongsamuth and Doran, 1997), and nearly 20 recombinant proteins, including reporter proteins (e.g. β-glucuronidase (GUS) and green fluorescent protein (GFP)) (Medina-Bolivar and Cramer, 2004; Lee *et al.*, 2007), enzymes (Gaume *et al.*, 2003; Woods *et al.*, 2008), mouse antibodies IgG anti-TMV (Sharp and Doran, 2001; Martinez *et al.*, 2005), antigens (Ko *et al.*, 2006; Rukavtsova *et al.*, 2007), growth factors (Komarnytsky *et al.*, 2006; Parsons *et al.*, 2010), immunomodulators such as ricin-B (the non-toxic lectin subunit of ricin as antigen carrier) (Medina-Bolivar *et al.*, 2003), hepatitis B surface antigens (HBsAg, in potato) (Kumar *et al.*, 2006), HBsAg (Rukavtsova *et al.*, 2007) and interleukin-12 (Liu *et al.*, 2009), have been expressed by hairy roots with protein yields up to 3.3% of total soluble protein (TSP) (Woods *et al.*, 2008). A company, ROOTec Bioactives Ltd (http://www.rootec.com), is currently working on commercialization of products with hairy roots grown in a mist bioreactor. Hairy roots might be an attractive *in vitro* expression system with several advantages compared to field-grown plants and suspension cultured cells. Future research will likely be focused on increasing expression levels and establishing effective and economical bioreactors for industrial production (Xu *et al.*, 2012).

3.2.3 Utilization of moss for recombinant proteins

The moss *Physcomitrella patens* is another promising platform in which photobioreactors are used for producing recombinant products. Knockout strains of *P. patens* lacking xylosylation and fucosylation activity have been produced so that recombinant proteins can have human-like glycosylation patterns (Koprivova *et al.*, 2004). Furthermore, humanized galactosylation patterns can be enhanced by adding human galactosyltransferase to the xylosyl and fucosyltranferase locus (Huether *et al.*, 2005). Using the moss as a platform, a variety of different engineered proteins have now been produced, including IgG1 and IgG4 antibodies, human VEGF, erythropoietin and factor H (Decker and Reski, 2008; Buttner-Mainik *et al.*, 2011) are a few recently cited examples. Greenovation (http://www.greenovation.com) has developed proprietary strains of *P. patens* and process technology for contract production of recombinant proteins up to 200-l scale using tubular or disposable rocker bag photobioreactors.

3.2.4 *Lemna* culture

Common duckweed or *Lemna minor* is an aquatic higher plant (*Lemnaceae* family), which is described as the world's smallest, fastest-growing and simplest flowering plant (Zhang *et al.*, 2010). *Lemna* has been identified for the production of recombinant proteins (Yamamoto *et al.*, 2001) because it is fast growing (doubling time about a day and half), simple to grow using photosynthesis, efficient to harvest (it generally floats on the liquid surface so can be skimmed off) and has up to 45% protein content (dry weight) (Stomp, 2005). Furthermore, *Lemna* is edible and considered to be a safe and alternative system for oral vaccines (Rival *et al.*, 2008). More than 20 therapeutic proteins including plasminogen (Spencer *et al.*, 2010), aprotinin (Rival *et al.*, 2008), monoclonal antibody (Cox *et al.*, 2006), avian influenza H5N1 haemagglutinin (Guo *et al.*, 2009) and interferon α2 (De Leede *et al.*, 2008) have been produced in duckweeds with expression levels up to 7% of TSP (Stomp, 2005). A reporter protein gene, *GFP*, expressed in *Spirodela oligorrhiza* reached a protein yield of more than 25% of TSP. This is among the highest expressing systems for nuclear transformation in a higher plant (Vunsh *et al.*, 2007). For the duckweed platform to be economically viable for industrial production, several issues need to be addressed including improved protein expression levels, human glycosylation, more genomic information on duckweed and duckweed reproduction, and methods for scale-up to commercial levels in both closed and open systems (Stomp, 2005). There are at least two products that are in preclinical development including a human plasmin (BLX-155) that is an anticoagulant, and an anti-CD20 mAb (BLX-301) (Paul and Ma, 2011; Maxmen, 2012).

3.2.5 Microalgae

Microalgae are another potential platform for cost-effective production of recombinant proteins. Microalgae are capable of rapid growth like microbes and can be cultured photoautotrophically in simple media while at the same time performing post-translational modifications found in higher plants (Potvin and Zhang, 2010; Specht *et al.*, 2010). Although studied for over a decade (Franklin and Mayfield, 2005), production of recombinant proteins in microalgae has been delayed over other systems such as bacteria and yeasts (Specht *et al.*, 2010), but lately the interest in microalgae as a viable bioproduction system has been renewed due to multiple factors (Rasala *et al.*, 2010; Specht *et al.*, 2010). The majority of current work is performed with the well-characterized green unicellular alga *Chlamydomonas reinhardtii* (Potvin and Zhang, 2010; Specht *et al.*, 2010). *Chlamydomonas* is an excellent system for reasons beyond its genetic and metabolic flexibility. This alga has a rapid doubling time (about 10 h), is easily scaled in homogeneous culture as an aqueous microbe, can be grown either photoautotrophically or with acetate as a reduced carbon source, and has a controllable and rapid sexual cycle (about 2 weeks) with stable and viable haploids (Specht *et al.*, 2010). Nuclear and chloroplast genomes of *C. reinhardtii* have been successfully transformed to express bioactive recombinant proteins. However, nuclear transformants generally failed to accumulate a high level of target molecules in comparison to chloroplast transformants, possibly due to nuclear silencing mechanisms (Specht *et al.*, 2010). As such, currently the chloroplast system is considered as the most feasible option for commercial production (Potvin and Zhang, 2010). A range of recombinant proteins has been successfully expressed in microalgae as documented in many excellent publications (Cadoret *et al.*, 2008; Potvin and Zhang, 2010; Specht *et al.*, 2010). These recombinant proteins include antibody (Xu *et al.*, 2011), a vaccine (Streatfield, 2006; Surzycki *et al.*, 2009), a blood protein (Manuell *et al.*, 2007), a growth factor (Rasala *et al.*, 2010) and an industrial enzyme (Yoon *et al.*, 2011). Finally, many species of green algae are considered GRAS (generally regarded as safe) (Rosenberg *et al.*, 2008), meaning that if the protein can be expressed in a bioavailable form, purification steps could potentially be eliminated altogether (Specht *et al.*, 2010). Although the proof of concept is established, recombinant protein production with microalgae is still in its infancy (Xu *et al.*, 2012). Systematic and concerted research efforts that are both biological- and engineering-based such as optimization of promoters, regulatory elements and codon usage as well as development of improved photobioreactor and product recovery systems will be critical to the success of the microalgal production platform.

Applications that directly use whole or minimally processed plants or plant parts are also being explored for industrial/bioenergy applications as well as therapeutics and vaccines to reduce further recombinant protein costs (Hood, 2002; Boothe *et al.*, 2010; Howard *et al.*, 2011). Plant-based production systems involving field-grown GM plants require regulatory oversight and approvals that are unique to plants compared to other bioproduction systems (Obembe *et al.*, 2011); regulatory considerations also impact production strategies and costs.

3.3 Commercial Production of Recombinant Proteins from Plant Cell Culture

The first plant-produced commercial therapeutic protein for human infusion, glucocerebrosidase, is produced in carrot cells by Protalix Biotherapeutics (http://www.protalix.com) and the drug has been approved by the US Food and Drug Administration (FDA). Currently, Dow AgroSciences (Indianapolis, Indiana; http://www.dowagro.com) is focused on the development and commercialization of pharmaceutical proteins expressed by proprietary plant cultures, named ProCellEx™ and Concert™ Plant-Cell-Produced System, respectively. Although recently liquidated, Biolex, Inc. developed the duckweed (*Lemna*)-based expression (LEX) system for producing interferon α2b (Locteron®) for the treatment of hepatitis C, which offers tolerability and dosing advantages over PEGylated interferons currently on the market (please see Section 3.7, p. 33 for further detail). Another company, Medicago (http://www.medicago.com) recently reported positive results from a Phase II clinical trial with its avian flu H5 pandemic vaccine candidate. The vaccine was found to be safe and well tolerated. It has been produced by using Medicago's Proficia™ technology, which is a vaccine and antibody production system based on transient expression of virus-like particles in plant leaves. The commercial successes of these proteins will bring a new era in the biopharmaceutical industry that promises to provide an affordable medicine produced from plant cells. Some other protein commercialization efforts using plants can be found in Maxmen (2012) and Broz et al. (2013).

3.4 Process Economics

Costs associated with processing for any commercial product are largely dependent on the final product. Thus, the final production cost will be the driving force for commercialization of plant-made recombinant proteins. The cost will be dependent on whether it is intended for an oral or skin-care therapeutic, a nutraceutical, a pharmaceutical or an industrial application. The product value will be much less for a functional food than to prepare a high purity pharmaceutical (Wilken and Nikolov, 2012). It is very important to keep the overall integration of process operation in mind during selection and process development of the product. Early analysis of developed processes is pivotal in transforming an R&D process into a manufacturing one (Nandi et al., 2005). This has an immense cost impact if processes are frozen at the early stage of clinical trial lot production (Rathore et al., 2004). The manufacturing cost for plant-produced proteins consists of upstream (biomass production) and downstream recovery and purification costs. The cost of manufacturing in most cases is a strong function of protein expression, overall process yield and production scale (Nandi et al., 2005).

The downstream processing costs are also affected by the ease of product recovery, the complexity of clarified plant or cell culture extracts, protein stability and required purity (Nikolov and Hammes, 2002). For example, biopharmaceuticals and processing enzymes for current Good Manufacturing Practices (cGMP) may require protein purities as high as 95–99% and those for diagnostics about 90%. Although in both of these cases the downstream manufacturing processes have to be robust (batch-to-batch repeatability), the main difference would be required documentation and regulatory-related activities. These are often 'hidden costs' in the biopharmaceutical manufacturing industry that are not readily available in the published literature and often either unaccounted for or underestimated (Farid, 2007). A scientific study using a discrete event modelling (DEM) approach reported that the projected cost of purified recombinant lactoferrin from rice seeds was US$5.90 g^{-1} (Nandi et al., 2005). It has been recently reported that cost-of-goods for the same cGMP grade product can be comfortably achieved at US$3.75 g^{-1} (Broz et al., 2013). This validated the idea that incorporating a developed methodology into the manufacturing process will have a major impact later in the process economics as long as the procedure has linear scalability of each step. The support activities, such as process and cleaning validation, buffer preparation, equipment cleaning, and quality control and quality assurance (QC/QA), can be a substantial fraction of operating costs. For example, the labour cost for validation and QC/QA activities can easily account for more than 50% of the direct manufacturing labour cost (Nikolov, personal communication). The breakdown of upstream production and downstream purification cost depends primarily on end application with biopharmaceutical and industrial proteins being at the opposite ends of the cost and purity spectrum (high to low). In general, the upstream cost for highly purified proteins (90% and above) from seed crops ranges from 5 to 10% of the total manufacturing cost depending on expression level, purification yield and annual product output (Evangelista *et al.*, 1998; Mison and Curling, 2000; Nikolov and Hammes, 2002; Nandi *et al.*, 2005).

The production cost (US$ g^{-1} product) of similar recombinant proteins produced in a contained system (greenhouse) would be three to five times greater than that produced in an open-field system (Pogue *et al.*, 2010). Therefore, downstream processing contribution to the total manufacturing cost of goods could range from 65 to 95%. There is less information and certainty about the cost of bioreactor-based systems, but one should anticipate 30–80 times greater production costs compared to the open field. Upstream production costs for low-purity protein products, such as industrial enzymes and oral vaccines, would be around 50% of the total manufacturing cost since minimal downstream processing would be required, as less costly recovery methods such as membrane filtration and protein precipitation are typically used (Nikolov and Hammes, 2002; Arntzen *et al.*, 2006). With regard to overall cost savings, the manufacturing cost for an open-field produced recombinant protein would be 30–50% lower than the cost of a bioreactor-produced protein, provided downstream processing costs were similar. The reduced regulatory costs associated with 'contained' processes may partly outweigh the currently lower productivities and higher capital costs of plant cell bioreactor-based platforms.

It is likely that the best way to achieve cost savings for a bioreactor-based system would be to maximize biomass productivity (including retaining and reusing cells in a semi-continuous or perfusion operation rather than batch culture), higher expression or product concentration, and downstream processing efficiency. Another important consideration for the production systems is the capital investment, which is rarely addressed in the published literature. By some estimates, capital investment increases multi-fold and proportionally to the level of containment and sophistication of plant growth controls with bioreactor-based systems approaching that of mammalian culture bioreactors (Spök and Karner, 2008). Thus, a decision-making process for choosing a plant production system is complex and requires case-by-case analysis. From a downstream cost perspective, several evaluation criteria could be applied for selecting the best system that matches product characteristics and allows overall manufacturing cost reduction (Nikolov and Hammes, 2002).

3.5 Regulatory Aspects

Increased concerns about regulatory compliance and product safety of mammalian systems have recently renewed interest in plant cell cultures as an alternative production platform for complex pharmaceutical proteins (Huang and McDonald, 2009; Shih and Doran, 2009; Xu et al., 2011). The above-mentioned systems provide a cGMP-compatible production environment more acceptable to the established pharmaceutical industry with added benefits of complex protein processing compared to bacteria and yeasts, and increased safety compared to mammalian cell systems which can harbour and propagate human pathogens.

Although proof-of-concept was originally established years ago using tobacco plants (Cramer et al., 1999; Radin et al., 1999; Conley et al., 2011), commercialization efforts were only started recently and are gathering momentum as described in the above section. Plant cell suspension culture combines the advantages of whole plants with those of microbial cultures (reviewed in Hellwig et al., 2004; Huang and McDonald, 2009; Xu et al., 2011). Although plant cell cultures do not share the benefit of unlimited scalability of field-cultivated whole plants, culture of plant cells in a sterile environment allows for precise control over growth and production conditions, batch-to-batch product consistency, flexibility in terms of ability to use inducible and plant viral-based expression systems, and a production process that is aligned with cGMP. The associated 'containment' aspects of both production and product reduce the regulatory burden compared to field-grown plants and provides a platform more consistent with established biopharmaceutical production systems, potentially aiding in both regulatory and industry acceptance. A suitable example of a reduced regulatory hurdle is the production of recombinant protein in microalgae. Many species of green algae (microalgae) are considered GRAS (Rosenberg et al., 2008), meaning that if the protein can be expressed in a bioavailable form, purification steps could potentially be eliminated or significantly reduced.

To attain a reasonable profit margin, the productivity of plant cell cultures needs to increase 10–50-fold, and this requires a systematic strategy to maximize the efficiency of all stages of the production pipeline from gene expression to cell culture, process development, and finally downstream protein purification (Weathers et al., 2010; Xu et al., 2011). Although whole-plant produced pharmaceuticals do not yet account for a significant portion of the preclinical and clinical pipeline, plant cells and tissues are emerging as a more compliant alternative 'factory' (Xu et al., 2011, 2012).

3.6 Choice of Different Bioreactor Systems for Plant Cell Culture

In the biopharmaceutical industry, a good technology portfolio, strong intellectual property position and access to capital might not guarantee success. Flexibility, cost effectiveness and time to market are key issues as well. Biopharmaceutical companies are keen on getting their products to market as quickly as possible to attract a majority of the possible market share. Therefore, the decision for future expansion of any product development process becomes impeded, as this decision must be made quite early, during product development stage. Such decisions are difficult to change later primarily due to regulatory constraints. To achieve an acceptable return on investment, biopharmaceutical companies focus on cutting down the cost of drug development and improving the overall time-to-market. For both development and final product from a cell culture process, the bioreactor plays a key role. Basic bioreactor designs and their utilization to produce a wide range of products, from antibiotics to foods to fuels, have been discussed by Williams (2002). Another recent review covers many important aspects of solid-state fermentation (Gutiérrez-Correa and Villena, 2012).

Nowadays safety, efficiency, time-saving capacity, reduced infrastructure requirements and productivity are the five major components that favour disposable bioreactor systems. Disposable bioreactors have been increasingly incorporated into preclinical, clinical and production-scale biotechnological facilities over the last few years (Eibl *et al.*, 2010). Single-use bags and components eliminate the risk of cross-contamination, reduce need of assembly and sterilization, cleaning and validation, and ultimately impact productivity. Moreover, use of disposable equipment also allows for quick changeover between products, which is valuable in the clinical phase of development, when often multiple products are evaluated simultaneously. Flexibility in using different operating strategies also helps increase productivity. A key factor in determining speed to market of disposables-based processes is associated with a timely decision to build the manufacturing facility. There are multiple options now available for different types of disposable bioreactors. A membrane-based system for high density cell culture, the CELLineTM bioreactor (http://www.integra-biosciences.com), has been developed for the production of monoclonal antibodies and other recombinant proteins (Trebak *et al.*, 1999; McDonald *et al.*, 2005; Adam *et al.*, 2008). In this system, a 'cell compartment' where the cells are grown is separated by a 10 kDa semi-permeable membrane from an upper 'nutrient compartment' where medium can be regularly replaced. Disposable bioreactors for cell cultures using wave-induced agitation (Singh, 1999) have also been considered as another option. This work describes a novel bioreactor system for the cultivation of animal, insect and plant cells using wave agitation induced by a rocking motion. This agitation system provides good nutrient distribution, off-bottom suspension and excellent oxygen transfer without damaging fluid shear or gas bubbles. The advantages of disposable bioreactors such as high flexibility, easy handling, reduced incidence of cross-contamination and savings in time and costs (Lim and Sinclair, 2007; Foulon *et al.*, 2008; Behme, 2009; Mauter, 2009) are attributed to the pre-sterility of the cultivation container, which is guaranteed by the vendor. The variety of disposable bioreactors currently available, encompassing wave-mixed, orbitally shaken or stirred reactors, and used to cultivate cells from a millilitre to cubic metre scale, have been reviewed (Eibl *et al.*, 2010; also see Table 1 in Eibl *et al.* that summarizes the next-generation of disposable bioreactors).

Despite the above advantages, the critical issues currently restricting the use of disposable bioreactors arise from the limited experience of using such bioreactors, concerns related to plastic material strength, leachables and scalability as well as the single-use philosophy itself (Eibl and Eibl, 2008). Renewal of the disposable cultivation container also contributes to an increase in the costs of solid waste disposal and consumables, resulting in higher running costs. Naturally, the cultivation task (biomass or cell production, expression of a biologically active substance) and the production of cell line, characterized by its morphology, growth and production behaviour, have a strong impact on the selection of the bioreactor type. Thus, the selection of a bioreactor for cell culture is dependent on a number of scientific and economic factors.

3.7 Glycoengineering in Plant Systems

Currently, many recombinant molecules are expressed in bacterial cell culture systems that are known to be cost effective, scalable and scientifically well understood, while allowing for fast, high-level expression of proteins. However, in many cases microbial systems often fail to deliver correctly folded and functional proteins (Wurm, 2004; Ozturk and Hu, 2006). In contrast, eukaryotic cells, including plant cells, exhibit a major advantage of allowing for the correct assembly and folding of recombinant polypeptides. Most proteins are indeed glycosylated, and these glycosylations involving many branched or linear chains, exhibiting particular O- or N-linkage (Delehedde *et al.*, 2006; Ohtsubo and Marth, 2006) have consequently made recombinant proteins more complex products to engineer than initially thought. Lately, there have been many examples of therapeutic products that have failed in clinics because they were not bearing the appropriate, if any, glycosylation (Harcum, 2006; Zucca *et al.*, 2006). Post-translational modifications are critical and are usually required for biological activity (Wurm, 2004; Kiss *et al.*, 2010). Experiments with sialylated proteins have demonstrated an ability to improve protein half-life in animal models. Studies involving recombinant human erythropoietins (rhEPO), where a sialylated version of the target protein continued to accumulate 9 days after infiltration when compared with a non-sialylated version that showed a gradual decrease in rhEPO over the same period of time (Jez *et al.*, 2013). Another study in mice involving recombinant butyrylcholinesterase (rBuChE) demonstrated that the polysialylated version of the protein had up to a sixfold increase in pharmacokinetic properties over the non-sialylated rBuChE, while providing a protection level virtually equal to that of the native version of the BuChE protein (Ilyushin *et al.*, 2013).

Production of therapeutically important proteins in plant cells has attracted increasing attention and initiated scientific investigation (Langer, 2010). Nevertheless, a barrier for producing human glycoproteins in plant cells is a significant difference in their N-glycan structures. Both high-mannose type and complex-type N-glycans are common in plant glycoproteins (Kiss *et al.*, 2010). Different from the complex N-glycan structures in human glycoproteins, which present with or without an α1-6-linked core fucose, plant complex N-glycan structures may have an α1-3-linked core fucose. In addition, instead of a bisecting β1-4-linked N-acetylglucosamine (GlcNAc) in human complex N-glycans, plant complex N-glycans may have a bisecting β1-2-xylose. Furthermore, unlike common β1-4-linked β-D-N-acetylgalactosamine (GalNAc) structures with or without an additional β1-3-linked fucose (e.g. Lewis x-type structures) in human complex N-glycans, β1-3-linked Gal1-3GlcNAc structures with or without an additional α1-3-linked fucose (e.g. lacto-N-biose or Lewis a-type structures) (Yu *et al.*, 2010) exist in plant glycans. Lastly, plants do not have a biosynthetic pathway for adding terminal sialic acid residues, which are commonly found in human glycoproteins. The presence of α1-3-linked core fucose and bisecting α1-2-xylose in plant glycoproteins has caused unwanted immunogenicity and stimulated the production of human-like glycoproteins as therapeutics by *in vivo* metabolic engineering of N-glycan biosynthetic pathways (Yu *et al.*, 2006; Sugiarto *et al.*, 2011), *in vitro* glycan remodelling using the combination of glycosidases and glycosyltransferases, or the combination of both methods. In recent years, most attempts to sialylate proteins that resemble native glycan structures have been carried out using plants and bacteria, with chemical and enzymatic modification systems using both *in vitro* and *in vivo* methods. The *in vitro* chemical modifications have primarily been done through PEGylation. In the process of PEGylation, a polyethylene glycol (PEG) chain is to be attached to a protein or peptide (Harris and Chess, 2003). Several studies have demonstrated the effectiveness of PEGylation in improving protein half-life in various animal models. PEGylated recombinant interleukin-11 (rhIL11) retention increased by about 60-fold over non-PEGylated rhIL11 in mice (Takagi *et al.*, 2007). In another study with nephrectomized rats (Zamboni, 2003), the rate of filgrastim cleared by the body decreased from 44.5 ml h^{-1} kg^{-1} in the non-PEGylated protein to 9.4 ml h^{-1} kg^{-1} in the PEGylated protein. A couple of recent studies showed site-specific enzymatic polysialylation of a therapeutic protein is possible (Lindhout *et al.*, 2011; Sohn *et al.*, 2013). Sometimes, simple sialylation reactions are not enough to increase the sialic acid content. A combined reaction using galactosyltransferase, sialyltrasferase and their sugar substrates at the same time is needed along with reduced incubation time to retain the activity while increasing sialylation (Sohn *et al.*, 2013). In recent developments at the University of California, Davis (Xi Chen, personal communication), they have developed several efficient one-pot multienzyme systems for adding GlcNAc or GalNAc, β1-3-linked galactose (Yu *et al.*, 2010), β1-4-linked galactose (Chen *et al.*, 2010), α1-3-linked fucose (Zhang *et al.*, 2010; Sugiarto *et al.*, 2011), as well as terminal α2-3- or α2-6-linked sialic acid (Yu *et al.*, 2005; Huang *et al.*, 2006; Sugiarto *et al.*, 2011; Thon *et al*, 2011a, b) to glycans and glycoconjugates. Therefore, it is expected that *in vitro* enzymatic glycan modification should be commercially achievable for recombinant proteins in the near future.

3.8 How Can Plant Cell Culture Systems Benefit Developing Countries?

Currently most of the therapeutics, drugs, diagnostic molecules, antibodies and vaccines are made of recombinant proteins. Costs of pharmaceuticals are increasing along with the global inflation, and in turn, half of the global population cannot keep up with the cost of medicine. So a clear need of continuous scientific improvement through technological intervention is unavoidable to reduce the cost of medicine significantly and make it available to most of the growing population, if not all. Health care delivery in the developing world is tied directly to the social and political will, or the extent of government engagement in the execution of health agendas and policies. Specifically, community-based governing bodies are the primary enforcers of government programmes and policies to improve the health of the local population (Langridge et al., 2012).

Plants are becoming commercially acceptable as recombinant protein production platforms for human therapeutics (Langer, 2010), vaccine antigens (Hefferon, 2013), industrial enzymes (Broz et al., 2013) and nutraceuticals (Maxmen, 2012). Many of these products will soon complement conventional pharmaceuticals in the treatment, prevention and diagnosis of disease, while at the same time adding value to agriculture. Such competition can be accelerated by developing better tools for the efficient exploration of diverse and mutually interacting arrays of phytochemicals and for the manipulation of the plant's ability to synthesize natural products and complex proteins (Raskin et al., 2002). Significant advances in expression, protein glycosylation and gene-to-product development time have been achieved as described in above sections. Safety and regulatory concerns for plant cell culture systems have also been addressed by using contained systems to grow transgenic plant cells. However, upstream technological achievements have yet to be matched by downstream processing advancements. In the past decade, the most research progress was achieved in areas of extraction and pre-treatment (Wilken and Nikolov, 2012). Many plant cells are considered as GRAS, as the final products need not be highly purified. Recently, non-chromatographic purification methods, such as aqueous two-phase partitioning and membrane filtration, have been evaluated as low cost purification alternatives to packed-bed adsorption (Wilken and Nikolov, 2012). Hefferon (2013) has discussed how and why plants offer tremendous advantages as cost-effective, safe and efficacious platforms for the large-scale production of vaccines and other therapeutic proteins. Plant-derived vaccines provide a way by which to enhance vaccine coverage for children in developing countries, and have the potential via oral administration to elicit a mucosal immune response (Hamorsky et al., 2013). Plants have the added advantage of simultaneously acting as an antigen delivery vehicle to the mucosal immune system while preventing the antigen from degradation as it passes through the gastrointestinal tract (Langridge et al., 2012).

While multiple deadly infectious diseases are almost eradicated in developed countries, they are still responsible for a high number of deaths in developing or underdeveloped countries (http://www.who.int/topics/ infectious_diseases/en). Recent advancement in science and technology in developed and some developing countries can successfully be translated to address this problem in an amicable way. An emphasis should particularly be given to diseases that have been eradicated or reduced significantly in developed countries, but are responsible for severe outbreaks in many developing countries. Using significant improvement in communication technology allows us to mobilize expertise and technology transfer in a much faster and efficient way than that of even a decade ago. In many cases using current infrastructure, knowledge and human resources that exist in some of the rapidly developing countries, this cost-effective technology can be implemented to develop affordable medicines (Langer, 2010; Langridge *et al.*, 2012; Hefferon, 2013). Thus plant cell cultures provide a new way of thinking about how some of our worst diseases can be dealt with, which will potentially benefit many of the world's underprivileged.

References

Adam, E., Sarrazin, S., Landolfi, C., Motte, V., Lortat-Jacob, H., Lassalle, P. and Delehedde, M. (2008) Efficient long-term and high-yielded production of a recombinant proteoglycan in eukaryotic HEK293 cells using a membrane-based bioreactor. *Biochemical Biophysical Research Communications* 369, 297–302.

Arntzen, C., Mahoney, R., Elliott, A., Holtz, B., Krattiger, A., Lee, C.K., *et al.* (2006) Plant-derived vaccines: cost of production. *Tempe: The Biodesign Institute at Arizona State University*.

Behme, S. (2009) Production facilities. In: Behme, S. (ed.) *Manufacturing of Pharmaceutical Proteins*. Wiley VCH, Weinheim, pp. 227–275.

Boothe, J., Nykiforuk, C., Shen, Y., Zaplachinski, S., Szarka, S., Kuhlman, P., Murry, E., Morck, D. and Moloney, M.M. (2010) Seed-based expression systems for plant molecular farming. *Plant Biotechnology Journal* 8, 588–606.

Broz, A., Huang, N. and Unruh, G. (2013) Plant-based protein biomanufacturing. *Genetic Engineering and Biotechnology News* 33.

Buttner-Mainik, A.P.J., Jerome, H., Hartmann, A., Lamer, S., Schaaf, A., Schlosser, A., Zipfel, P.F., Reski, R. and Decker, E.L. (2011) Production of biologically active recombinant human factor H in *Physcomitrella*. *Plant Biotechnology Journal* 9, 373–383.

Cadoret, J.P.B.M., Lerouge, P., Cabigliera, M., Henriquez, V. and Carlier, A. (2008) Microalgae as cell factories producing recombinant commercial proteins. *Medical Science (Paris)* 24, 375–382.

Chen, J., Huang, S., Yu, H., Li, Y., Lau, K. and Chen, X. (2010) Trans-sialidase activity of *Photobacterium damsela* alpha2,6-sialyltransferase and its application in the synthesis of sialosides. *Glycobiology* 20 260–268.

Conley, A.J., Zhu, H., Le, L.C., Jevnikar, A.M., Lee, B.H., Brandle, J.E. and Menassa, R. (2011) Recombinant protein production in a variety of *Nicotiana* hosts: a comparative analysis. *Plant Biotechnology Journal* 9, 419–433.

Cox, K.M., Sterling, J.D., Regan, J.T., Gasdaska, J.R., Frantz, K.K., Peele, C.G., Black, A., Passmore, D., Moldovan-Loomis, C., Srinivasan, M., Cuison, S., Cardarelli, P.M. and Dickey, L.F. (2006) Glycan optimization of a human monoclonal antibody in the aquatic plant *Lemna minor*. *Nature Biotechnology* 24, 1591–1597.

Cramer, C.L., Boothe, J.G. and Oishi, K.K. (1999) Transgenic plants for therapeutic proteins: linking upstream and downstream strategies. *Current Topics in Microbiology and Immunology* 240, 95–118.

Daniell, T. and Edwards, R. (1995) Changes in protein methylation associated with the elicitation response in cell-cultures of alfalfa (*Medicago sativa* L). *FEBS Letters* 360, 57–61.

De Leede, L.G., Humphries, J.E., Bechet, A.C., Van Hoogdalem, E.J., Verrijk, R. and Spencer, D.G. (2008) Novel controlled-release *Lemna*-derived IFN-alpha2b (Locteron): pharmacokinetics, pharmaco-dynamics, and tolerability in a phase I clinical trial. *Journal of Interferon Cytokine Research* 28, 113–122.

Decker, E.L. and Reski, R. (2008) Current achievements in the production of complex biopharmaceuticals with moss bioreactors. *Bioprocessing and Biosystem Engineering* 31, 3–9.

Delehedde, M., Sarrazin, S., Adam, E., Motte, V. and Vanpouille, C. (2006) Proteoglycans and glycosaminoglycans: complex molecules with modulating activity. In: Delehedde, M. (ed.) *New Developments in Therapeutic Glycomics*. Research Signpost, Kerala, pp. 1–13.

Eibl, R. and Eibl, D. (2008) Bioreactors for mammalian cells: general overview. In: Eibl, R., Eibl, D., Pörtner, R., Catapano, G. and Czermak, P. (eds) *Cell and Tissue Reaction Engineering*. Springer, Heidelberg, pp. 55–82.

Eibl, R., Kaiser, S., Lombriser, R. and Eibl, D. (2010) Disposable bioreactors: the current state-of-the-art and recommended applications in biotechnology. *Applied Microbiology and Biotechnology* 86, 41–49.

Evangelista, R.L., Kusnadi, A.R., Howard, J.A. and Nikolov, Z.L. (1998) Process and economic evaluation of the extraction and purification of recombinant β-Glucuronidase from transgenic corn. *Biotechnology Progress* 14, 607–614.

Farid, S.S. (2007) Process economics of industrial monoclonal antibody manufacture. *Journal of Chromatography B* 848, 8–18.

Foulon, A., Trach, F., Pralong, A., Proctor, M. and Lim, J. (2008) Using disposables in an antibody production process: a cost-effectiveness study of technology transfer between two production sites. *BioProcess* 6, 12–18.

Franklin, S.E. and Mayfield, S.P. (2005) Recent developments in the production of human therapeutic proteins in eukaryotic algae. *Expert Opinion on Biological Therapy* 5, 225–235.

Gaume, A., Komarnytsky, S., Borisjuk, N. and Raskin, I. (2003) Rhizosecretion of recombinant proteins from plant hairy roots. *Plant Cell Reports* 21, 1188–1193.

Giri, A. and Narasu, M.L. (2000) Transgenic hairy roots: recent trends and applications. *Biotechnology Advances* 18, 1–22.

Guillon, S., Tremouillaux-Guiller, J., Pati, N.K., Rideau, M. and Gantet, P. (2006) Harnessing the potential of hairy roots: dawn of a new era. *Trends in Biotechnology* 24, 403-409.

Guo, X., Bublot, M., Pritchard, N., Dickey, L., Thomas, C. and Swayne, D.E. (2009) *Lemna* (duckweed) expressed hemagglutinin from avian influenza H5N1 protects chickens against H5N1 high pathogenicity avian influenza virus challenge. *Abstracts of the 7th International Symposium on Avian Influenza, 2009, Athens, Georgia*, 62.

Gutiérrez-Correa, M. and Villena, G.K. (2012) Batch and repeated batch cellulase production by mixed cultures of *Trichoderma reesei* and *Aspergillus niger* or *Aspergillus phoenicis*. *Journal of Microbiology and Biotechnology Research* 2, 929–935.

Hamorsky, K.T., Kouokam, J.C., Bennett, L.J., Baldauf, K.J., Kajiura, H., Fujiyama, K. and Matoba, N. (2013) Rapid and scalable plant-based production of a cholera toxin B subunit variant to aid in mass vaccination against cholera outbreaks. *PLoS Neglected Tropical Diseases* 7, e2046.

Harcum, S. (2006) Protein glycosylation. In: Ozturk, S. and Hu, W. (eds) *Cell Culture Technology for Pharmaceutical and Cell-Based Therapies*. Taylor & Francis, New York, pp. 113–154.

Harris, J.M. and Chess, R.B. (2003) Effect of pegylation on pharmaceuticals. *Nature Reviews, Drug Discovery* 2, 214–221.

Hefferon, K. (2013) Plant-derived pharmaceuticals for the developing world. *Biotechnology Journal (Special Issue: Plant Biotechnology)* 8, 1193–1202.

Hellwig, S., Drossard, J., Twyman, R.M. and Fischer, R. (2004) Plant cell cultures for the production of recombinant proteins. *Nature Biotechnology* 22, 1415–1422.

Hood, E.E. (2002) From green plants to industrial enzymes. *Enzyme and Microbial Technology* 30, 279–283.

Howard, J.A., Nikolov, Z. and Hood, E.E. (2011) Enzyme production systems for biomass conversion. In: Hood, E.E., Nelson, P. and Powell, R. (eds) *Plant Biomass Conversion*. Wiley Press, Ames, Iowa, pp. 227–253.

Huang, J., Sutliff, T.D., Wu, L., Nandi, S., Benge, K., Terashima, M., Ralston, A.H., Drohan, W., Huang, N. and Rodriguez, R.L. (2001) Expression and purification of functional human alpha-1-Antitrypsin from cultured plant cells. *Biotechnology Progress* 17, 126–133.

Huang, J.M., Wu, L.-Y., Yalda, D., Adkins, Y., Kelleher, S.L., Crane, M., Lonnerdal, B., Rodriguez, R.L. and Huang, N. (2002) Expression of functional recombinant human lysozyme in transgenic rice cell culture. *Transgenic Research* 11, 229–239.

Huang, L.F., Liu, Y.K., Lu, C.A., Hsieh, S.L. and Yu, S.M. (2005) Production of human serum albumin by sugar starvation induced promoter and rice cell culture. *Transgenic Research* 14, 569–581.

Huang, L., Shou, T., Chen, X., Yu, H., Sun, C. and Liang, Z. (2006) Slab like functional architecture of higher order cortical area 21a showing oblique effect of orientation preference in the cat. *Neuroimage* 32, 1365–1374.

Huang, T.K. and McDonald, K.A. (2009) Bioreactor engineering for recombinant protein production in plant cell suspension cultures. *Biochemical Engineering Journal* 45, 168–184.

Huether, C.M., Lienhart, O., Baur, A., Stemmer, C., Gorr, G., Reski, R. and Decker, E.L. (2005) Glyco-engineering of moss lacking plant-specific sugar residues. *Plant Biology* 7, 292–299.

Ilyushin, D.G., Smirnov, I.V., Belogurov, A.A., Dyachenko, I.A., Zharmukhamedova, T.I., Novozhilova, T.I. and Gabibov, A.G. (2013) Chemical polysialylation of human recombinant butyrylcholinesterase delivers a long-acting bioscavenger for nerve agents *in vivo*. *Proceedings of the National Academy of Sciences USA* 110, 1243–1248.

Jez, J., Castilho, A., Grass, J., Vorauer, U.K., Sterovsky, T., Altmann, F. and Steinkellner, H. (2013) Expression of functionally active sialylated human erythropoietin in plants. *Biotechnology Journal* 8, 371–382.

Jo, S.H., Kwon, S.Y., Park, D.S., Yang, K.S., Kim, J.W., Lee, K.T., Kwak, S.S. and Lee, H.S. (2006) High-yield production of functional human lactoferrin in transgenic cell cultures of Siberian ginseng (*Acanthopanax senticosus*). *Biotechnology and Bioprocess Engineering* 11, 442–448.

Kiss, Z., Elliott, S., Jedynasty, K., Tesar, V. and Szegedi, J. (2010) Discovery and basic pharmacology of erythropoiesis-stimulating agents (ESAs), including the hyperglycosylated ESA, darbepoetin alfa: an update of the rationale and clinical impact. *European Journal of Clinical Pharmacology* 66, 331–340.

Ko, S., Liu, J.R., Yamakawa, T. and Matsumoto, Y. (2006) Expression of the protective antigen (SpaA) in transgenic hairy roots of tobacco. *Plant Molecular Biology Reporter* 24.

Komarnytsky, S., Borisjuk, N., Yakoby, N., Garvey, A. and Raskin, I. (2006) Cosecretion of protease inhibitor stabilizes antibodies produced by plant roots. *Plant Physiology* 141, 1185–1193.

Koprivova, A., Stemmer, C., Altmann, F., Hoffmann, A., Kopriva, S., Gorr, G., Reski, R. and Decker, E.L. (2004) Targeted knockouts of *Physcomitrella* lacking plantspecific immunogenic N-glycans. *Plant Biotechnology Journal* 2, 517–523.

Kumar, G.B.S., Ganapathi, T.R., Srinivas, L., Revathi, C.J and, Bapat, V.A. (2006) Expression of hepatitis B surface antigen in potato hairy roots. *Plant Science* 170, 918–925.

Kwon, T.H., Kim, Y.S., Lee, J.H. and Yang, M.S. (2003a) Production and secretion of biologically active human granulocyte-macrophage colony stimulating factor in transgenic tomato suspension cultures. *Biotechnology Letters* 25, 1571–1574.

Kwon, T.H., Seo, J.E., Kim, J., Lee, J.H., Jang, Y.S. and Yang, M.S. (2003b) Expression and secretion of the heterodimeric protein interleukin-12 in plant cell suspension culture. *Biotechnology and Bioengineering* 81, 870–875.

Langer, E.S. (2010) Plant expression systems growing rapidly: use of the technology for vaccine manufacture leads the way toward commercialization. *Genetic Engineering and Biotechnology News* 30.

Langridge, W., Odumosu, O., Nandi, S., Rodriguez, R., Deleon, M. and Cordero-Macintyre, Z. (2012) Mucosal vaccination against enteric pathogens in the developing world. *British Journal of Medicine & Medical Research* 2, 260–291.

Lee, K.T., Chen, S.C., Chiang, B.L. and Yamakawa, T. (2007) Heat-inducible production of beta-glucuronidase in tobacco hairy root cultures. *Applied Microbiology and Biotechnology* 73, 1047–1053.

Lienard, D., Dinh, O.T., Van Oort, E., Van Overtvelt, L., Bonneau, C., Wambre, E., Bardor, M., Cosette, P., Didier-Laurent, A., de Borne, F.D., Delon, R., van Ree, R., Moingeon, P., Faye, L. and Gomord, V. (2007) Suspension-cultured BY-2 tobacco cells produce and mature immunologically active house dust mite allergens. *Plant Biotechnology Journal* 5, 93–108.

Lim, J.A.C. and Sinclair, A. (2007) Process economy of disposable manufacturing: process models to minimize upfront investment. *American Pharmaceutical Review* 10, 114–121.

Lindhout, T., Iqbal, U., Willis, L.M., Reid, A.N., Li, J., Liu, X. and Wakarchuk, W.W. (2011) Site-specific enzymatic polysialylation of therapeutic proteins using bacterial enzymes. *Proceedings of the National Academy of Sciences USA* 108, 7397–7402.

Liu, C.Z., Towler, M.J., Medrano, G., Cramer, C.L. and Weathers, P.J. (2009) Production of mouse interleukin-12 is greater in tobacco hairy roots grown in a mist reactor than in an airlift reactor. *Biotechnology and Bioengineering* 102, 1074–1086.

Magnuson, N.S., Linzmaier, P.M., Reeves, R., An, G.H., Hayglass, K. and Lee, J.M. (1998) Secretion of biologically active human interleukin-2 and interleukin-4 from genetically modified tobacco cells in suspension culture. *Protein Expression and Purification* 13, 45–52.

Manuell, A.L., Beligni, M.V., Elder, J.H., Siefker, D.T., Tran, M., Weber, A., McDonald, T.L. and Mayfield, S.P. (2007) Robust expression of a bioactive mammalian protein in *Chlamydomonas* chloroplast. *Plant Biotechnology Journal,* 5, 402–412.

Martinez, C., Petruccelli, S., Giulietti, A.M.A. and Alvarez, M.A. (2005) Expression of the antibody 14D9 in *Nicotiana tabacum* hairy roots. *Electronic Journal of Biotechnology* 8, 170–176.

Matsumoto, S., Ishii, A., Ikura, K., Ueda, M. and Sasaki, R. (1993) Expression of human erythropoietin in cultured tobacco cells. *Bioscience, Biotechnology and Biochemistry* 57, 1249–1252.

Mauter, M. (2009) Environmental life-cycle assessment of disposable bioreactors. *BioProcess* 7, 18–28.

Maxmen, A. (2012) Drug-making plant blooms: Approval of a 'biologic' manufactured in plant cells may pave the way for similar products. *Nature Biotechnology* 485, 160.

McDonald, K.A., Hong, L.M., Trombly, D.M., Xie, Q. and Jackman, A.P. (2005) Production of human alpha-1-antitrypsin from transgenic rice cell culture in a membrane bioreactor. *Biotechnology Progress* 21, 728–734.

Medina-Bolivar, F. and Cramer, C. (2004) Production of recombinant proteins by hairy roots cultured in plastic sleeve bioreactors. *Methods in Molecular Biology* 267, 351–363.

Medina-Bolivar, F., Wright, R., Funk, V., Sentz, D., Barroso, L., Wilkins, T.D., Petri, W. and Cramer, C.L. (2003) A non-toxic lectin for antigen delivery of plant-based mucosal vaccines. *Vaccine* 21, 997–1005.

Mison, D. and Curling, J. (2000) The industrial production costs of recombinant therapeutic proteins expressed in transgenic corn. *Biopharm International* 13, 48–54.

Nandi, S., Yalda, D., Lu, S., Nikolov, Z., Misaki, R., Fujiyama, K. and Huang, N. (2005) Process development and economic evaluation of recombinant human lactoferrin expressed in rice grain. *Transgenic Research* 14, 237–249.

Nikolov, Z.L. and Hammes, D. (2002) Production of recombinant proteins from transgenic crops. In: Hood, E.E. and Howard, J.A. (eds) *Plants as Factories for Protein Production.* Kluwer Academic Publishers, Dordrecht, the Netherlands, pp. 159–174.

Obembe, O.O., Popoola, J.O., Leelavathi, S. and Reddy, S.V. (2011) Advances in plant molecular farming. *Biotechnology Advances* 29, 210–222.

Ohtsubo, K. and Marth, J.D. (2006) Glycosylation in cellular mechanisms of health and disease. *Cell* 126, 855–867.

Ozawa, K. and Takaiwa, F. (2010) Highly efficient *Agrobacterium*-mediated transformation of suspension-cultured cell clusters of rice (*Oryza sativa* L.). *Plant Science* 179, 333–337.

Ozturk, S.S. and Hu, W. (2006) Cell culture technology – an overview. In: Ozturk, S.S. and Hu, W. (eds) *Cell Culture Technology for Pharmaceutical and Cell-Based Therapies.* Taylor & Francis, New York.

Parsons, J., Wirth, S., Dominguez, M., Bravo-Almonacid, F., Giulietti, A.M. and Rodriguez, T.J. (2010) Production of human epidermal growth factor (hEGF) by *in vitro* cultures of *Nicotiana tabacum*: effect of tissue differentiation and sodium nitroprusside addition. *International Journal of Biotechnology and Biochemistry* 6, 131–138.

Paul, M. and Ma, J.K.C. (2011) Plant-made pharmaceuticals: leading products and production platforms. *Biotechnology and Applied Biochemistry* 58, 58–67.

Pogue, G.P., Vojdani, F., Palmer, K.E., Hiatt, E., Hume, S., Phelps, J., Long, L., Borohova, N., Kim, D., Pauly, M., Velasco, J., Whaley, K., Zeitlin, L., Garger, S.J., White, E., Bai, Y., Haydon, H. and Bratcher, B. (2010) Production of pharmaceutical-grade recombinant aprotinin and a monoclonal antibody product using plant-based transient expression systems. *Plant Biotechnology Journal* 8, 638–654.

Potvin, G. and Zhang, Z.S. (2010) Strategies for high-level recombinant protein expression in transgenic microalgae: a review. *Biotechnology Advances* 28, 910–918.

Radin, D.N., Cramer, C.L., Oishi, K.K. and Weissenborn, D.L. (1999) Production of Human Lysosomal Proteins in Plant-based Expression Systems. United States patent application.

Rasala, B.A., Muto, M., Lee, P.A., Jager, M., Cardoso, R.M.F., Behnke, C.A., Kirk, P., Hokanson, C.A., Crea, R., Mendez, M. and Mayfield, S.P. (2010) Production of therapeutic proteins in algae, analysis of expression of seven human proteins in the chloroplast of *Chlamydomonas reinhardtii*. *Plant Biotechnology Journal* 8, 719–733.

Raskin, I., Ribnicky, D.M., Komarnytsky, S., Ilic, N., Poulev, A., Borisjuk, N., Brinker, A., Moreno, D.A., Ripoll, C., Yakoby, N., O'Neal, J.M., Cornwell, T., Pastor, I. and Fridlender, B. (2002) Plants and human health in the twenty-first century. *Trends in Biotechnology* 20, 522–531.

Rathore, A.S., Latham, P., Levine, H., Curling, J. and Kaltenbrunner, O. (2004) Costing issues in the production of biopharmaceuticals. *Biopharmocology International* 17, 46–55.

Rival, S., Wisniewski, J.P., Langlais, A., Kaplan, H., Freyssinet, G., Vancanneyt, G., Vunsh, R., Perl, A. and Edelman, M. (2008) *Spirodela* (duckweed) as an alternative production system for pharmaceuticals: a case study, aprotirin. *Transgenic Research* 17, 503–513.

Rosenberg, J.N., Oyler, G.A., Wilkinson, L. and Betenbaugh, M.J. (2008) A green light for engineered algae: redirecting metabolism to fuel a biotechnology revolution. *Current Opinion in Biotechnology* 19, 430–436.

Rukavtsova, E.B., Abramikhina, T.V., Shulga, N.Y., Bykov, V.A. and Bur'yanov, Y.I. (2007) Tissue specific expression of hepatitis B virus surface antigen in Transgenic plant cells and tissue culture. *Russian Journal of Plant Physiology* 54, 770–775.

Shaaltiel, Y., Bartfeld, D., Hashmueli, S., Baum, G., Brill-Almon, E., Galili, G., Dym, O., Boldin-Adamsky, S.A., Silman, I., Sussman, J.L., Futerman, A.H. and Aviezer, D. (2007) Production of glucocerebrosidase with terminal mannose glycans for enzyme replacement therapy of Gaucher's disease using a plant cell system. *Plant Biotechnology Journal* 5, 579–590.

Shanks, J.V. and Morgan, J. (1999) Plant 'hairy root' culture. *Current Opinion in Biotechnology* 10, 151–155.

Sharma, A.K. and Sharma, M.K. (2009) Plants as bioreactors: recent developments and emerging opportunities. *Biotechnology Advances* 27, 811–832.

Sharp, J.M. and Doran, P.M. (2001) Strategies for enhancing monoclonal antibody accumulation in plant cell and organ cultures. *Biotechnology Progress* 17, 979–992.

Shih, S.M.H. and Doran, P.M. (2009) Foreign protein production using plant cell and organ cultures: advantages and limitations. *Biotechnology Advances* 27, 1036–1042.

Shin, Y.J., Hong, S.Y., Kwon, T.H., Jang, Y.S. and Yang, M.S. (2003) High level of expression of recombinant human granulocyte-macrophage colony stimulating factor in transgenic rice cell suspension culture. *Biotechnology and Bioengineering* 82, 778–783.

Shin, Y.J., Lee, N.J., Kim, J., An, X.H., Yang, M.S. and Kwon, T.H. (2010) High-level production of bioactive heterodimeric protein human interleukin-12 in rice. *Enzyme Microbial Technology* 46, 347–351.

Sijmons, P.C., Dekker, B.M., Schrammeijer, B., Verwoerd, T.C., Van Den Elzen, P.J. and Hoekema, A. (1990) Production of correctly processed human serum albumin in transgenic plants. *Nature Biotechnology* 8, 217–221.

Singh, V. (1999) Disposable bioreactor for cell culture using wave-induced motion. *Cytotechnology* 30, 149–158.

Smith, M.L., Mason, H.S. and Shuler, M.L. (2002) Hepatitis B surface antigen (HBsAg) expression in plant cell culture: kinetics of antigen accumulation in batch culture and its intracellular form. *Biotechnology and Bioengineering* 80, 812–822.

Sohn, Y., Lee, J.M., Park, H.-R, Jung, S.-C, Park, T.H. and Oh, D.-B. (2013) Enhanced sialylation and *in vivo* efficacy of recombinant human α-galactosidase through *in vitro* glycosylation. *BMB Reports* 46, 157–162.

Specht, E., Miyake-Stoner, S. and Mayfield, S. (2010) Micro-algae come of age as a platform for recombinant protein production. *Biotechnology Letters* 32, 1373–1383.

Spencer, D., Dickey, L.F., Gasdaska, J.R., Wang, X., Cox, K.M. and Peele, C.G. (2010) Expression of plasminogen and microplasminogen in duck weed. United States patent application.

Spök, A. and Karner, S. (2008) Plant molecular farming: opportunities and challenges. In: Stein, A.J. (ed.) *The Institute for Prospective Technological Studies*. European Commission, Seville.

Stomp, A.M. (2005) The duckweeds: a valuable plant for biomanufacturing. *Biotechnology Annual Review* 11, 69–99.

Streatfield, S.J. (2006) Engineered chloroplasts as vaccine factories to combat bioterrorism. *Trends in Biotechnology* 24, 339–342.

Su, W.W. and Lee, K.T. (2007) Plant cell and hairy-root cultures-process characteristics, products, and application. In: Yang, S.T. (ed.) *Bioprocessing for Value-added Products from Renewable Resources*. Elsevier Science, Amsterdam, the Netherlands, pp. 263–292.

Sugiarto, G., Lau, K., Yu, H., Vuong, S., Thon, V., Li, Y., Huang, S. and Chen, X. (2011) Cloning and characterization of a viral alpha2-3-sialyltransferase (ST3Gal) for the synthesis of sialyl Lewis[x]. *Glycobiology* 21, 387–396.

Surzycki, R., Greenham, K., Kitayama, K., Dibal, F., Wagner, R., Rochaix, J.D., Ajam, T. and Surzycki, S. (2009) Factors effecting expression of vaccines in microalgae. *Biologicals* 37, 133–138.

Takagi, A., Yamashita, N., Yoshioka, T., Takaishi, Y., Sano, K., Yamaguchi, H. and Hashida, M. (2007) Enhanced pharmacological activity of recombinant human interleukin-11 (rhIL11) by chemical modification with polyethylene glycol. *Journal of Controlled Release* 119, 271–278.

Thon, V., Li, Y., Yu, H., Lau, K. and Chen, X. (2011a) PmST3 from *Pasteurella multocida* encoded by Pm1174 gene is a monofunctional alpha 2-3-sialyltransferase. *Applied Microbiology and Biotechnology* 94(4), 977–985.

Thon, V., Lau, K., Yu, H., Tran, B. K. and Chen, X. (2011b) PmST2: a novel *Pasteurella multocida* glycolipid alpha2-3-sialyltransferase. *Glycobiology* 21, 1206–1216.

Torres, E., Vaquero, C., Nicholson, L., Sack, M., Stoger, E., Drossard, J., Christou, P., Fischer, R. and Perrin, Y. (1999) Rice cell culture as an alternative production system for functional diagnostic and therapeutic antibodies. *Transgenic Research* 8, 441–449.

Trebak, M., Chong, J.M., Herlyn, D. and Speicher, D.W. (1999) Efficient laboratory-scale production of monoclonal antibodies using membrane-based high density cell culture technology. *Journal of Immunology Methods* 230, 59–70.

Tremblay, R., Wang, D., Jevnikar, A.M. and Ma, S. (2010) Tobacco, a highly efficient green bioreactor for production of therapeutic proteins. *Biotechnology Advances* 28, 214–221.

Trexler, M.M., Mcdonald, K.A. and Jackman, A.P. (2005) A cyclical semicontinuous process for production of human alpha(1)-antitrypsin using metabolically induced plant cell suspension cultures. *Biotechnology Progress* 21, 321–328.

Vunsh, R., Li, J.H., Hanania, U., Edelman, M., Flaishman, M., Perl, A., Wisniewski, J.P. and Freyssinet, G. (2007) High expression of transgene protein in Spirodela. *Plant Cell Reports* 26, 1511–1519.

Weathers, P.J., Towler, M.J. and Xu, J.F. (2010) Bench to batch: advances in plant cell culture for producing useful products. *Applied Microbiology and Biotechnology* 85, 1339–1351.

Wilken, L.R. and Nikolov, Z.L. (2012) Recovery and purification of plant-made recombinant proteins. *Biotechnology Advances* 30, 419–433.

Williams, J.A. (2002) Keys to bioreactor selections. *CEP Magazine*, 34–41.

Wongsamuth, R. and Doran, P.M. (1997) Production of monoclonal antibodies by tobacco hairy roots. *Biotechnology and Bioengineering* 54, 401–415.

Woods, R.R., Geyer, B.C. and Mor, T.S. (2008) Hairy-root organ cultures for the production of human acetylcholinesterase. *BMC Biotechnology* 8.

Wurm, F.M. (2004) Production of recombinant protein therapeutics in cultivated mammalian cells. *Nature Biotechnology* 22, 1393–1398.

Xu, J., Dolan, M.C., Medrano, G., Cramer, C.L. and Weathers, P.J. (2012) Green factory: Plants as bioproduction platforms for recombinant proteins. *Biotechnology Advances* 30, 1171–1184.

Xu, J.F., Ge, X.M. and Dolan, M.C. (2011) Towards high-yield production of pharmaceutical proteins with plant cell suspension cultures. *Biotechnology Advances* 29, 278–299.

Yamamoto, Y.T., Rajbhandari, N., Lin, X.H., Bergmann, B.A., Nishimura, Y. and Stomp, A.M. (2001) Genetic transformation of duckweed *Lemna gibba* and *Lemna minor. In Vitro Cellular and Devopmental Biology, Plant* 37, 349–353.

Yoon, S.M., Kim, S.Y., Li K.F., Yoon, B.H., Choe, S. and Kuo, M.M. (2011) Transgenic microalgae expressing *Escherichia coli* AppA phytase as feed additive to reduce phytate excretion in the manure of young broiler chicks. *Applied Microbiology and Biotechnology* 91, 553–563.

Yu, H., Chokhawala, H., Karpel, R., Wu, B., Zhang, J., Zhang, Y., Jia, Q. and Chen, X. (2005) A multifunctional *Pasteurella multocida* sialyltransferase: a powerful tool for the synthesis of sialoside libraries. *Journal of the American Chemical Society* 127, 17618–17619.

Yu, H., Chokhawala, H.A., Huang, S. and Chen, X. (2006) One pot three enzyme chemoenzymatic approach to the synthesis of sialosides containing natural and non-natural functionalities. *Nature Protocols* 1, 2485–2492.

Yu, H., Thon, V., Lau, K., Cai, L., Chen, Y., Mu, S., Li, Y., Wang, P.G. and Chen, X. (2010) Highly efficient chemoenzymatic synthesis of beta1-3-linked galactosides. *Chemical Communications (Cambridge)* 46, 7507–7509.

Zamboni, W.C. (2003) 'Pharmacokinetics of pegfilgrastim' Pharmacotherapy. *The Journal of Human Pharmacology and Drug Therapy* 23, 9–14.

Zhang, Y., Hu, Y., Yang, B., Ma, F., Lu, P., Li, L., Wan, C., Rayner, S. and Chen, S. (2010) Duckweed (*Lemna minor*) as a model plant system for the study of human microbial pathogenesis. *PLoS One* 5, e13527.

Zucca, A., Brizzi, S., Riccioni, R., Azzara, A., Ghimenti, M. and Carulli, G. (2006) Glycosylated and nonglycosylated recombinant human granulocyte colony-stimulating factor differently modifies actin polymerization in neutrophils. *La Clinica Terapeutica* 157, 19–24.

4 Plant-derived Monoclonal Antibodies as Human Biologics for Infectious Disease and Cancer

Qiang Chen* and Huafang Lai

The Biodesign Institute and School of Life Sciences, Arizona State University, Arizona

4.1 Introduction

Current human biologics, including those based on monoclonal antibodies (MAbs), are commonly produced by fermentation technologies using primarily mammalian cell cultures. However, its high cost and low scalability severely limit this platform from meeting the ever-increasing global demand. Plants offer a novel system for the development and production of biologics that is more scalable, cost-effective, speedy, versatile and safer than current expression paradigms. The possibility of producing low-cost human biologics on an agricultural scale is extremely attractive for commercial biologics production, as well as for manufacturing pharmaceuticals for the developing world. This chapter focuses on the importance of plants as an innovative, speedy, flexible and economical system for developing and producing MAb-based human biologics. The recent development of deconstructed virus-based vectors that have allowed rapid and high-level transient expression of MAbs in plants is first presented. The progress in plant glycoengineering that allows plants to produce MAbs with superior efficacy and safety than other traditional systems is subsequently described. The combined advantages of these new breakthroughs, which have promoted the plant expression system to become a premier platform for the commercial development and production of human biologics, are extensively discussed in context of several leading MAb-based biologics.

4.2 Plants as a Source of Natural Pharmaceuticals

Plants have served as an important source for traditional medicines for many centuries. Many traditional cultures around the globe use whole plants or plant parts as herbs to treat a variety of conditions. Secondary metabolites from plant extracts have also been widely used to treat wound

* Corresponding author, E-mail: Qiang.Chen.4@asu.edu

and microbial infections. Even today, plants still provide more than a quarter of prescription drugs on the market (Farnsworth *et al.*, 1985; Duke, 1993). One of the most popular examples of these natural plant-derived pharmaceuticals is morphine, the painkiller alkaloid from opium plants. In addition to single-molecule drugs like morphine, plants also offer natural pharmaceuticals in the form of 'botanicals', a mixture of functionally synergistic molecules for treating many diseases (Chen, 2011a). The high demand and broad applications for these natural plant pharmaceuticals utilized throughout human history hint at the major advantages of plants as a system for producing human pharmaceuticals. First is the safety that plants have exhibited as a source for human pharmaceuticals. Unlike animal or animal cells, plants generally are not subject to infection by nor carry human pathogens. Correspondingly, plant pathogens such as plant viruses do not infect humans. As a result, the risk of introducing human or animal pathogens to humans by plant-derived pharmaceuticals is greatly reduced (Chen, 2011b). Second is the large capacity plants have demonstrated in providing pharmaceuticals. Initially, the traditional pharmaceuticals were extracted from wild plants. As their demand increased, these plants were cultivated first in relatively small plots of land, and subsequently in larger areas. For plants rare in nature or hard to cultivate, plant tissue culture was invented to replace whole plant material. The scale-up process for plant cultivation and plant tissue culture is relatively simpler than that of livestock or mammalian cell culture. Lastly, the traditional use of plant pharmaceuticals has also demonstrated the low-cost nature of plants as a production system. In comparison to raising animal or establishing mammalian cell culture facilities, the upstream production of plant materials is much cheaper and can be managed through routine agricultural practices, requiring only land, water, minerals and the sun. This economical advantage is reflected in both the initial setup and the subsequent scale-up stages, leading to significant cost savings in capital investment for building large facilities and in operational costs. The cost of downstream processing for recovering pharmaceuticals from plant tissue has evidently been low enough for these pharmaceuticals to have been widely used in the developing world. Thus, plants have been successfully exploited as a production source of human pharmaceuticals for a long time. The demonstrated safe, low-cost and scalable nature of these pharmaceuticals highlight plants as a superb system for developing and producing the next generation of human biologics.

4.3 Plants as a Novel System for Development and Production of Protein-based Biologics

Protein-based biologics are the future of medicine. This new class of molecules is the result of continued innovations by scientists in finding cures for life-threatening diseases. Most biologics today are produced in mammalian cell cultures (Yin *et al.*, 2007). This popular system has been optimized for many years and matured into a remarkable platform for the production of biologics. However, cell culture-based systems, including mammalian cells, have inherent challenges, such as prohibitive cost, limited scalability and the risk of carrying human or animal pathogens. Building a mammalian cell culture facility requires heavy up-front capital investment and a long lead time for production (Hiatt and Pauly, 2006). Operational costs are also significantly higher than other expression systems as it needs expensive culture media and precise electronic control of growth conditions. Furthermore, construction of duplicated facilities and fermentation tanks is necessary to accommodate larger-scale production, creating challenges in scalability. In addition, the risk of contamination of mammalian cell-derived biologics with human or animal pathogens is a constant public health safety concern. These challenges may prevent the full realization of the vast potential of biologics and call for the development of new production platforms that are robust, low-cost, scalable and safe.

Plants have demonstrated their potential in overcoming the cost, scalability and safety challenges in producing natural plant pharmaceuticals (Chen, 2008, 2011a). The advancement of molecular biology has permitted the use of these advantages for producing modern biologics. Specifically, the development of DNA recombinant techniques has enabled the introduction of transgenes into plant cells, allowing them to produce non-native recombinant biologics (Lico et al., 2008). Since the first production of a recombinant human growth hormone in plants in 1986 (Barta et al., 1986), a broad range of functionally active biologics has been produced by a variety of plant expression systems. These successes indicate the great potential of plants as a platform for human biologics production. In addition to the traditional advantages of low cost, high scalability and increased safety, plant expression systems also offer other benefits. For example, plants share the eukaryotic endomembrane system with human cells. Consequently, in contrast to bacterium-based systems, plant cells can efficiently perform protein post-translational modifications and assembly of multiple subunits that are required for the functional activity of most human biologics (Chen, 2008; Li and da'Anjou, 2009).

Biologics can be produced in plants via one of three major strategies, including: (i) using stable plant lines with a transformed nuclear genome; (ii) plants with a transformed chloroplast genome; and (iii) plants transiently expressing the transgene (Chen, 2011a). For the first two strategies, the gene of interest is cloned into an expression vector and delivered into plant cells. Transformed cells with a transgene stably integrated into their nuclear or chloroplast genome are selected for generating transgenic plants. This process can take several months to a year. However, once accomplished, the Mendelian inheritance of the transgene is achieved, allowing the stable expression of the biologic over many generations. Thus, the first two strategies can lead to the propagation and stock-up of master seed banks for large-scale production of biologics.

To create stable transgenic plants, transgene expressing cassettes are introduced into plant cells either by direct biolistics or indirect delivery through *Agrobacterium tumefaciens* (Chen et al., 2013). In biolistics (also known as microprojectile bombardment) methods, a transgene cassette is coated on to gold or tungsten particles and ballistically shot into plant cells (Sanford, 1990). Indirect gene delivery methods rely on *Agrobacterium* species that naturally transfer DNA into plant cells. As a physical process, biolistic delivery through a gene gun can be applied to cells of virtually any plant species and can deliver transgenes to both nuclear and chloroplast genomes. Biolistic delivery usually results in a random integration of the transgene, often with multiple copies found throughout the genome. In contrast, plant transformation with *A. tumefaciens* is limited by the natural host range of the bacterium, namely most dicotyledonous and a few monocotyledonous plant species (Klee *et al.*, 1987). However, it has been shown that *A. tumefaciens*-based methods outperform biolistics significantly in transformation efficiency, transgene expression and inheritance (Rivera *et al.*, 2012). These differences may be due to the more selective integration of the transgene into genomic areas with more active transcriptional activities, as a result of the co-evolution of *Agrobacterium* and its plant hosts (Klee *et al.*, 1987).

The third strategy of producing biologics in plants is transient expression. In contrast to the first two strategies, the transgene is not integrated into the plant genomes. Instead, the transgene is actively transcribed and translated while being in the cell transiently (Komarova *et al.*, 2010). Thus, transient expression accelerates the production speed by eliminating the often tedious and slow process of transgenic plant generation and selection. Transient expression also enhances the level of transgene transcription by eliminating position effects due to the location of transgene integration. The development of expression vectors based on plant viruses has further enhanced the yield of recombinant biologics due to their robustness in replication, transcription and translation (Lico *et al.*, 2008). Several different types of vectors including replacement, fusion and insertion vectors have been developed. For example, a cauliflower mosaic virus (CaMV)-based replacement vector was the first viral vector that successfully demonstrated the expression of a recombinant protein in plants (Brisson *et al.*, 1984). Replacement vectors based on tomato bushy stunt virus (TBSV) and tobacco mosaic virus (TMV) have also been used in expressing a variety of transgenes (Lico *et al.*, 2008). Examples of fusion vectors include the fusion of vaccine epitopes on to the viral coat protein (CP) so that they are displayed on the surface of the virus (Porta *et al.*, 2003; Lico *et al.*, 2006). In insertion vectors, including those derived from TMV and potato virus X (PVX), a transgene is inserted into a complete functional viral genome, and is expressed as a by-product of the viral genome replication cycle (Scholthof *et al.*, 1993; Musiychuk *et al.*, 2007). Overall, plant viral vectors allow high-level accumulation of biologics due to the robust transcription and translation of plant viruses. However, whole viral-based vectors do suffer from several drawbacks. For instance, vectors based on double-stranded DNA viruses have very small packaging capacity. They often lose essential functions even when a small fraction of their genome is replaced by a transgene (Brisson *et al.*, 1984). CP is usually the target of gene substitution for replacement vectors. Since it is essential for the cell-to-cell movement of many viruses, CP replacement often leads to the loss of systemic infectivity. In turn, this reduces the number of cells to which the transgene is delivered. In fact, delivery of viral vectors to plant cells is generally problematic. While mechanical inoculation of infectious viral particles or viral nucleic acids is feasible for many viruses, other viruses are not susceptible to mechanical rubbing but require a specialized insect for their transmission. Moreover, a cumbersome and unscalable *in vitro* process of RNA vector generation is required for RNA viruses. The narrow host range of plant viruses further hinders the broad application of the whole virus-based vectors in transient expression.

The new development of 'deconstructed' viral vectors is a promising breakthrough. It has not only effectively overcome the limitations of whole viral vectors, but also further increased the transgene expression in transient systems. Deconstruction is achieved by deleting viral genome components that are not essential or beneficial for the vector function. As a result, the size of the viral replicon is significantly reduced, which leaves a larger capacity to accommodate transgene insertion. Furthermore, the deconstructed viral genome can be delivered into plant cells in the form of DNA or cDNA. For vectors derived from RNA viruses, autonomous replicons are produced when the DNA construct is transcribed and spliced in plant cells. This feature eliminates the unscalable *in vitro* transcription process of generating RNA vectors. It has also allowed the use of *A. tumefaciens* for the delivery of transgene in transient expression systems, which provides numerous advantages (Chen *et al.*, 2013). For example, the need for viral systemic infection is eliminated as *A. tumefaciens* can deliver vectors to most of the cells on the entire plant (Leuzinger *et al.*, 2013). Consequently, CP can be deleted to accommodate large sizes of transgenes. Furthermore, transgene loss during systemic spreading is no longer a concern. Delivery with *A. tumefaciens* also broadens the range of suitable plant hosts because it can transfer DNA to cells that are not susceptible to mechanical infection. Overall, deconstructed viral vectors effectively integrate the benefits of three biological systems. First, it retains the speed and high protein yield of the whole viral systems, but eliminates the risk of creating infectious viral particles due to the removal of CP genes. Second, it fully utilizes the ability of *A. tumefaciens* in delivering transgenes into plant cells. Consequently, these vectors are more robust, scalable and can be applied to more plant host species, because CP and the *in vitro* process of creating RNA constructs are no longer needed. Finally, the post-translational processing capacity of eukaryotic plant cells for assembling and modifying complex multi-subunit proteins is integrated into the system. Thus, expression systems based on deconstructed viral vectors gain the flexibility of nuclear gene expression with the speed and expression amplification of viral vectors.

One of the widely used deconstructed vectors is the MagnICON system, based on TMV and PVX (Marillonnet *et al.*, 2004; Giritch *et al.*, 2006). It is a system with three pro-vector cDNA modules, with the 5′ module carrying the viral RNA-dependent RNA polymerase, the 3′ module containing the transgene, and a third module coding for a recombination integrase (Giritch *et al.*, 2006). Once delivered and expressed in plant cells, the integrase assembles the 5′ and 3′ modules into a replication-competent TMV or PVX genome under the control of a plant promoter. This assembled construct is then transcribed into a functional replicon. The results from our laboratory and others have collectively demonstrated that this system can drive a very high level of accumulation of biologics of various sizes, from small subunit vaccines to large immune complexes (Giritch *et al.*, 2006; Santi *et al.*, 2008; Chen *et al.*, 2009; Lai *et al.*, 2010; Phoolcharoen *et al.*, 2011a; He *et al.*, 2012; Lai and Chen, 2012). Moreover, the high level expression can be achieved rapidly within 7–10 days after agro-infiltration.

The geminiviral DNA replicon system based on bean yellow dwarf virus (BeYDV) is another prominent example (Chen *et al.*, 2011). We developed this system to overcome the challenge of other deconstructed viral vectors in producing multiple hetero-subunit protein biologics. For vectors derived from many viruses, co-delivery of a mix of vectors with the same viral components often results in segregation and preferential amplification of only one vector in a single cell, a phenomenon called 'competing replicons' (Hull and Plaskitt, 1970; Dietrich and Maiss, 2003). For example, TMV and PVX are both competing viruses, but not with each other. This allows the MagnICON system to produce biologics with two hetero-subunits, such as monoclonal antibodies (MAbs) with the heavy (HC) and light chain (LC) gene built on the TMV and PVX backbone separately. As a result, both HC and LC can be expressed in the same cell, permitting their proper assembly into a functional MAb (Giritch *et al.*, 2006). However, biologics with three or more distinct subunits, such as secretory IgA, IgM and certain viral-like particle (VLP) vaccine candidates, cannot be produced by the MagnICON system currently (Latham and Galarza, 2001; Chen and Lai, 2013; Thuenemann *et al.*, 2013). Furthermore, the prospect of identifying additional viruses that are compatible with both TMV and PVX is not encouraging. We have circumvented these problems by developing a non-competing vector system based on BeYDV. This plant virus belongs to the *Geminiviridae* family. One of the important characteristics of BeYDV is the ability to replicate its single-stranded circular DNA genome to very high copy numbers by a rolling circle mechanism upon infection of plants (Liu *et al.*, 1997). Interestingly, only two *cis*-acting elements (the long intergenic region (LIR) and the short intergenic region (SIR)) and a single viral protein (replication associated protein (Rep)) are required for viral replication (Chen *et al.*, 2011). Correspondingly, we developed the BeYDV geminiviral vectors with the transgene cassette inserted between the LIR and the SIR (Huang *et al.*, 2009; Chen *et al.*, 2011). In the first generation of vectors, we supplied the gene for the Rep protein in a second module (Huang *et al.*, 2009). Our studies indicated that this two-module system can drive high-level accumulation of biologics in plants through a robust replication of the transgene-carrying replicon (Huang *et al.*, 2009). We optimized this system in the second-generation vectors by integrating the transgene and Rep modules into a single vector system. This improvement greatly simplifies the upstream process for biologic production particularly for large-scale operations (Huang *et al.*, 2010). Most importantly, we demonstrated that the BeYDV geminiviral system is non-competing and allows for the efficient expression and assembly of multi-subunit biologics, such as MAbs (Huang *et al.*, 2010). For this purpose, we further optimized the system by creating a single vector that contains multiple replicon cassettes, each encoding for a different protein or subunit. We demonstrated that upon delivery into plant cells, each cassette in this single vector assembles into an independent replicon and produces high levels of the protein/subunit it codes for, without interfering with the replication of other replicons or the production of other proteins/subunits in the same plant cell (Huang *et al.*, 2010). This is a speedy expression system as the highest accumulation of biologics is usually within 4–8 days post-agroinfiltration, slightly shorter than the MagnICON system (Huang *et al.*, 2009, 2010; Lai *et al.*, 2012). The plant host range may also be broader than the MagnICON system as demonstrated by studies of our group (Lai *et al.*, 2012).

Overall, the geminiviral replicon system allows for the production of biologics with more than two hetero-subunits, while retaining the speed and high production capacity of other deconstructed viral systems. Moreover, it is more scalable as the need for co-infiltration of multiple expression modules is obviated. In addition to the geminiviral and MagnICON systems, other deconstructed viral vector systems, including those based on 5′ and 3′ untranslated regions (UTRs) of cowpea mosaic virus (CPMV) RNA-2 and tobacco yellow dwarf mastrevirus (TYDV), have been developed and show great promise as robust vectors for transient expression (Sainsbury and Lomonossoff, 2008; Sainsbury *et al.*, 2010; Dugdale *et al.*, 2013). Collectively, the development of deconstructed viral vectors marks a significant advancement in transient expression technology and positions it as a competitive platform for commercial production of biologics.

All three plant expression strategies have potential applications for the development and production of biologics. Transgenic plants with a stably integrated transgene provide a seed bank – a permanent genetic source for the cost-effective propagation of plants and commercial scale production of biologics. However, the difficult and time-consuming process of establishing a high-quality seed bank severely hinders the application of this technology. These difficulties are derived from technical factors including the long lead time required for generating transgenic plants, the fluctuation of gene expression levels among individual plants and between plant generations, and potential transgene silencing (Chen, 2011b). Public concern over genetically modified (GM) plants further complicates its current application. In spite of these challenges, stable transgenic plants provide the most scalable technology for very large-scale commercial production of biologics. Plants with a stably transformed chloroplast genome produce significant higher level of recombinant proteins than nuclear transgenic plants due to the sheer large number of chloroplasts in each cell (Daniell, 2006). Transplastomic plants also offer other benefits in addressing both the technical and regulatory challenges facing transgenic plant technology. For example, the 'position effect' of nuclear transformation, which causes inconsistent level of transgene expression, can be eliminated by targeting the transgene to specific positions of the chloroplast genome through homologous recombination. The maternal transmission of chloroplasts in most plants reduces the risk of transgene escape through pollen, making this strategy a preferred alternative for ensuring transgene containment and addressing public concerns for GM plants. However, the niche for this strategy is limited to biologics that do not require post-translational modifications for their function due to lack of post-translational modification pathways in chloroplasts.

Thus, the transient, deconstructed viral-based expression systems present the most promising plant-based strategy for the development and production of biologics, with their focus on production speed, yield and flexibility (Hiatt and Pauly, 2006). Due to the rapid and high yield accumulation of biologics, the initial application of this platform is to develop new biologics and/or obtain the initial material (at a milligram to gram level) for pre-clinical or phase I trial characterization. Besides the speed and high yield of the transient expression system, an additional advantage is the versatility for producing personalized MAb-based therapeutics and vaccines against viruses that have rapid antigenic drift and/or multiple strains with unpredictable epidemics. This advantage also arms this type of transient system with the 'surge' capability to rapidly produce biologics to address a bioterror event. Since this strategy does not produce stable transgenic plants or an intact replication-competent plant virus, the risk of transgene spreading through the routes of pollen, seeds or virus is eliminated, making it a favourable choice from a public acceptance and regulatory perspective. This strategy, however, is not limited to research and development or small-scale production. Since transient expression systems use non-transgenic materials for vector delivery, plant biomass can be readily produced by conventional agricultural practice for large-scale commercial production without generating the concerns of GM plants. Furthermore, new spray-based *Agrobacterium* delivery technologies are being developed to deliver target genes into field-grown plants, which will allow the application of transient expression across large agricultural scales (He *et al.*, 2014a). In fact, some of the new expression vectors already combine the strengths of both the stable and transient expression systems, allowing the stable inheritance of the transgene and the robust, yet controlled transient expression of biologics with a specific chemical signal (Dugdale *et al.*, 2013). These 'bridge' vectors offer a complete platform for the rapid assessment of biologic candidates and their transition to a large-scale commercial production.

4.4 Plant-produced Monoclonal Antibody-based Biologics

4.4.1 Monoclonal antibodies

The development of MAbs has revolutionized the pharmaceutical industry and provided new opportunities for resolving a wide range of difficult medical problems. Mammalian cell culture-produced MAbs have achieved remarkable pharmaceutical and financial success. The current global market of human biologics is dominated by MAbs, accounting for over half of the total therapeutics produced, with the top ten earning more than US$52 billion in 2011 (Mullard, 2012). However, the high cost, long manufacturing time and limited capacity of the cell culture-based manufacture system severely hinders the availability of these drugs and the realization of their vast potential. These challenges can be overcome by using plant expression systems. Antibody production requires a eukaryotic host cell that can assemble the four antibody polypeptides into a heterotetramer and perform complex glycosylation. Despite this complexity, a MAb was successfully expressed in tobacco plants only 3 years after the first plant-made biologic (Hiatt et al., 1989). Since then, a variety of MAbs and their derivatives, such as secretory IgAs, single-domain fragments, single-chain variable fragments (scFv) and diabodies have been successfully produced in many plant species, with an increasing number and type of MAb being produced every year (Table 4.1). Early MAb production was performed in stable transgenic plants. As the case for other biologics, it suffered from low MAb yield, long lead time to generate and select stable transgenic plants, plant-specific glycosylation and unstable seed banks. These problems arose from the randomness of transgene integration in the plant genome (position effect), the lack of control in transgene copy number, the inconsistent ratio of LC and HC, and the shortage of strong promoters to drive transgene expression (Pogue et al., 2010). As a result, early research focused on resolving technical issues such as finding the optimal relative orientation of the LC and HC cassette rather than structural and functional analysis. These challenges greatly undermined the cost-saving potential of the plant expression systems. The development of transient expression systems based on deconstructed virus vectors has revolutionized MAb production in plants. Our group has extensively used the transient system in producing MAbs and their derivatives. We have demonstrated with the MagnICON and geminiviral vectors that MAbs can be routinely obtained within 10 days of vector infiltration into *Nicotiana benthamiana* plants with yield up to 1mg MAb g^{-1} fresh leaf weight. In addition to *Nicotiana* plants, deconstructed viral vectors also deliver rapid and high-level accumulation of MAbs in other host plants including commercially produced plant materials (Huang et al., 2010; Lai et al., 2010; Phoolcharoen et al., 2011a; He et al., 2012; Lai and Chen, 2012). This greatly enhances the feasibility of using this system for commercial production. With better yield, it is possible to shift the focus to a more detailed structural and functional analysis of plant-derived MAbs. For example, our group has published the first report that demonstrated the efficacy of a plant-produced MAb against a potentially lethal infection several days after exposure in an animal challenge model (Lai et al., 2010). Studies of several MAbs have reached the stage of human clinical trials and demonstrated their safety and efficacy (Table 4.2). Collectively, these studies demonstrated that plant-derived MAbs share similar structures and therapeutic activities with commercially licensed MAbs.

Table 4.1. Representative plant-produced MAbs, MAb-fragments and MAb fusion biologics.

Antibody type	Pharmaceutical target	Plant host/expression system	Efficacy in animal model or human	Sponsor/Reference
Fab	Neuron and rheumatic diseases	Stable transgenic *Arabidopsis thaliana*	Not reported	Peeters *et al.*, 2001
diabody	Diagnostic and therapeutic for HCG-expressing cancers or as contraceptive	*Nicotiana tabacum*/ transient expression with 35S-based vector	Not reported	Kathuria *et al.*, 2002
scFv, IgG1	Therapeutic vaccine for non-Hodgkin's lymphoma	*Nicotiana benthamiana*/ transient expression with Geneware, MagnICON	Protection in mice, safety in humans	Icon (McCormick *et al.*, 2003, 2008; Bendandi *et al.*, 2010)
Secretory IgA/G	Tooth decay	Stable transgenic *N. tabacum*	Protection against colonization by oral streptococci in humans	Planet Biotechnology (Ma *et al.*, 1998)
IgG1	Therapeutic for Ebola	*N. benthamiana*, lettuce/transient expression with geminiviral, MagnICON	Protection in mice and rhesus macaques	Mapp/KBP/ASU (Huang *et al.*, 2010; Zeitlin *et al.*, 2011; Olinger *et al.*, 2012)
RIC	Vaccine for Ebola	*N. benthamiana*/ transient expression with geminiviral, MagnICON	Protection in mice	ASU (Phoolcharoen *et al.*, 2011a,b)
scFv-Fc	Therapeutic for WNV	*N. benthamiana*/ transient expression with geminiviral, MagnICON	Post-exposure protection even after WNV entered into CNS in mice	ASU (Lai *et al.*, 2014)
IgG1	Therapeutic for WNV	*N. benthamiana*, lettuce/transient expression with geminiviral, MagnICON	Post-exposure protection even after WNV entered into CNS in mice	ASU (Lai *et al.*, 2010, 2012)
IgG1	Therapeutic for respiratory syncytial virus	*N. benthamiana*/ transient expression MagnICON	Prophylactic and therapeutic protection in cotton rats	Mapp (Zeitlin *et al.*, 2013)
IgG1	Breast cancer treatment	*N. benthamiana*/ transient expression with 35S-based, MagnICON	Reduction in tumour volume in mice	Icon (Komarova *et al.*, 2011)

Fab, antigen binding fragment; scFv, single-chain variable fragment; RIC, Recombinant Immune Complex; Fc, fragment crystallizable region; ASU, Arizona State University; KBP, Kentucky Bioprocessing; WNV, West Nile virus.

Table 4.2. Examples of plant-derived human biologics that have been approved by FDA or reached human clinical trial stage.

Product	Disease application	Plant host	Development stage
Glucocerebrosidase (ELELYSO™)	Gaucher's disease therapeutic enzyme	Carrot cell culture	Approved by FDA
Guy's 13 sIgA (CaroRx)	Prevent tooth decay	Transgenic tobacco	Phase II completed, approved for use in EU
ICAM-1 and sIgA fusion (RhinoRx)	Rhinovirus-related common cold	Transgenic tobacco	Phase II
scFvs	Tumour-specific personalized vaccine for non-Hodgkin's lymphoma	*Nicotiana benthamiana* infected with a full viral vector	Phase I completed
IgG1	Tumour-specific personalized vaccine for non-Hodgkin's lymphoma	*N. benthamiana* transient expression	Phase I
IgG1 (2G12)	HIV prophylactic	Transgenic tobacco	Phase I completed
Influenza A H5N1 HA enveloped virus-like particles	Pandemic flu vaccine	*N. benthamiana* transient expression	Phase II completed
Alpha Interferon (Locteron)	Hepatitis C therapeutic	Transgenic duckweeds	Phase IIb completed
Insulin	Diabetes therapeutic	Transgenic safflower	Phase I/II completed

sIgA, secretory immunoglobulin A; ICAM-1, intercellular adhesion molecule-1; HA, haemagglutinin; EU, European Union.

4.4.2 MAbs against flavivirus

Flavivirus belongs to a genus of viruses in the *Flaviviridae* family that includes West Nile virus (WNV), dengue virus (DV), Japanese and St Louis encephalitis virus, tick-borne encephalitis virus and yellow fever virus. The flaviviruses are enveloped viruses with a positive-sense single-stranded RNA genome. The viral genome replicates itself in the cytoplasm of the host cell and produces a viral polyprotein upon translation. The polyprotein is subsequently processed into several polypeptides by proteases of both host and viral origin. In general, these polypeptides mature into three structural proteins, i.e. the capsid protein (CP), the membrane (M) protein and the Envelope (E) protein, and eight non-structural proteins.

WNV is a neurotropic virus that infects the central nervous system (CNS) of humans and animals. Historically, WNV was an old world disease mostly found in the eastern hemisphere with distribution in Africa and Asia. However, since 1999, WNV entered the western hemisphere and quickly spread across the USA, Canada, the Caribbean region and Latin America (Hubalek and Halouzka, 1999). In the USA, the frequency and severity of WNV outbreaks have increased significantly in recent years with 2012 as the deadliest (286 fatalities) on record. Elderly people are the most vulnerable for developing severe neurological disease, long-term morbidity and death (Bode *et al.*, 2006). Studies also identified genetic factors for high susceptibility (Diamond and Klein, 2006; Lim *et al.*, 2008). Currently, there is no vaccine nor protein therapeutics approved for human use. The global threat of WNV epidemics and the lack of treatment warrant the development of protein therapeutics and production platforms that can expeditiously bring the products to the biologic market at low cost. In response to this challenge, our research group explored a plant-derived MAb (E16) as a therapeutic candidate. E16 is a humanized murine MAb that binds to an epitope on domain III (DIII) of WNV E protein. It was predicted that it would have significant therapeutic potential for WNV because it can disrupt virus transmission between neurons (Samuel *et al.*, 2007). However, the mammalian cell culture-produced E16 may be too costly for health-care systems to realize its therapeutic potential.

To ensure high-level expression in plants, we first optimized the coding sequence of E16 LC and HC *in silico* and cloned them into the 5' modules of plant expression vectors of the MagnICON system (Lai *et al.*, 2010). These vectors were then co-infiltrated into *N. benthamiana* leaves along with other components of the MagnICON system (Leuzinger *et al.*, 2013). Our results indicate the LC and HC of E16 were produced in leaves with the expected molecular weights and they assembled into the MAb tetrameric (2HC + 2LC) form. Our results also demonstrate that E16 was rapidly produced in leaves and accumulated at its highest level (0.81 mg g^{-1} leaf fresh weight (LFW)) 7 days post-infiltration (dpi) (Lai *et al.*, 2010).

In addition to producing E16 in *N. benthamiana* with MagnICON vectors, our research group also investigated the feasibility of expressing E16 with geminiviral vectors in lettuce, as well as in *N. benthamiana*. Results from this study showed that the geminiviral replicon system permits robust expression and assembly of E16 in both *N. benthamiana* and lettuce plants with accumulation levels comparable to those achieved with the MagnICON vectors in *N. benthamiana* (Lai *et al.*, 2012). Moreover, the geminiviral vectors drive more rapid expression of E16 and reached the highest accumulation level at 4 dpi, 3 days earlier than that of the MagnICON system. Overall, these results demonstrate that plants can rapidly express fully-assembled E16 at a level greater than the highest plant MAbs expression levels ever reported.

For E16 to become a viable WNV therapeutic, it has to be recovered and purified from the plant tissue efficiently. We developed a three-step purification scheme with ammonium sulfate precipitation, followed by protein A affinity and DEAE-anion exchange chromatography. It was shown that ammonium sulfate precipitation removed the most plant host proteins, including the photosynthetic enzyme RuBisCo (Fig. 4.1, Lane 2). The subsequent protein A affinity chromatography removed the remaining contaminating proteins. This two-step process enriched E16 to greater than 95% purity. Analyses with SDS-PAGE indicate that purified E16 appeared as the pure HC and LC in the expected stoichiometric ratio under reducing conditions (Fig. 4.1, Lanes 5 and 6) and was detected as the expected assembled tetrameric band in the presence of an oxidizing agent (Fig. 4.1, Lane 7). A third anion-exchange chromatography was included in our protocol for compliance with the current Good Manufacturing Practice

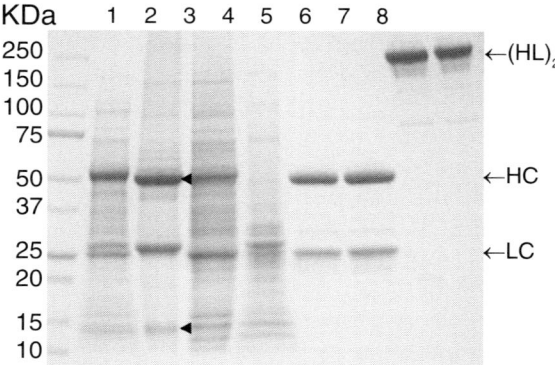

Fig. 4.1. Purification of E16 MAb from *N. benthamiana* plants (Lai *et al.*, 2010). Leaf protein extract was purified and analysed on a 4–20% SDS-PAGE gel under a reducing (lanes 1–6) or non-reducing (lanes 7 and 8) condition. Lane 1, clarified plant extract; lane 2, plant proteins removed by 25% ammonium sulfate precipitation; lane 3, 50% ammonium sulfate pellet fraction re-suspended for protein A chromatography; lane 4, protein A flow-through fraction; lanes 5 and 7, purified pHu-E16 mAb in the protein A eluate; lanes 6 and 8, mHu-E16 as a reference standard. ◄, RuBisCo large and small subunits; ←, LC, HC and assembled form (HL)$_2$ of Hu-E16 mAb.

(cGMP) regulations for future clinical material production. Our results demonstrate that contaminants and/or impurities, including residual DNA, endotoxin and protein A, were efficiently removed, and their levels in the final E16 product were below the Food and Drug Administration (FDA) specifications for injectable human MAb biologics (Table 4.3). In addition to *N. benthamiana*, we also used this processing scheme to purify E16 from lettuce. We discovered that since lettuce leaves produce negligible amounts of phenolics and alkaloids compared with *Nicotiana* plants, the ammonium sulfate precipitation step, which is partially responsible for removing secondary metabolites from tobacco extract, can be bypassed. Thus, lettuce clarified extracts can be directly loaded on to the protein A column without the concern for protein A fouling by phenolics and alkaloids (Lai *et al.*, 2012). The scalability of the purification scheme was also demonstrated, as E16 was purified from *N. benthamiana* plants with consistent purity and recovery among batches of different scale (Table 4.4).

The structural, biochemical and functional properties of plant-derived E16 were compared with its mammalian cell-produced counterpart. It was shown by both ELISA and flow cytometry analyses that there is no difference in the recognition of WNV DIII antigen between plant- and mammalian cell-derived E16. Furthermore, quantitative analysis by a surface plasmon resonance (SPR) assay demonstrates that plant- and mammalian cell-produced E16 have almost identical binding affinity and kinetics for WNV E protein and DIII (Lai *et al.*, 2010). The neutralization activity of E16 was evaluated by a quantitative flow cytometry assay, which showed equivalent neutralization of WNV infection between plant and mammalian E16 (Lai *et al.*, 2010). These *in vitro* functional studies suggested the efficacy of the

Table 4.3. Characterization of E16 MAb purified from plants (adapted from Lai *et al.*, 2010).

Plant host/batch no.	Residual DNA (ng ml^{-1})	Residual Protein A (ng ml^{-1})	Endotoxin (EU ml^{-1})	Compliance with FDA specifications
N. benthamiana #1	<1	9.77 ± 3.02	3.78 ± 1.52	Yes
N. benthamiana #2	<1	11.65 ± 2.15	3.57 ± 2.60	Yes
N. benthamiana #3	<1	12.04 ± 3.42	2.94 ± 1.57	Yes
N. benthamiana #4	<1	10.33 ± 6.65	4.12 ± 2.93	Yes
Lettuce	<1	13.54 ± 4.67	3.40 ± 1.83	Yes

Data (mean ± SD) from three independent batches for each batch are presented.

Table 4.4. Examples of MAb purification scheme scalability (adapted from Lai *et al.*, 2010).

LFW(g)	Recovery (%)	Purity
10	57.52 ± 2.59	> 95%
100	51.71 ± 2.86	> 95%
500	45.77 ± 4.84	> 95%
5000	48.76 ± 6.06	> 95%

LFW, leaf fresh weight. Data (mean ± SD) from three independent batches for each biomass scale.

plant-produced E16. To confirm this hypothesis *in vivo*, the potency of plant-produced E16 was tested in a mouse model. Prophylaxis studies were performed in wild type (WT) C57BL/6 mice to identify the concentrations of E16 that could prevent severe WNV infection. Mice were infected with a lethal dose (10^2 PFU) of WNV and on the same day administered a single dose of 0.001 to 10 µg plant- or mammalian cell-derived E16. We found that mice were significantly protected when injected with as little as 0.1 µg of plant-derived E16 (Fig. 4.2a, P <0.001). More than 80% of mice were prevented from lethal infection when 10 µg of Hu-E16 was administered (P <0.0001). Overall, plant-derived E16 prevented mice from WNV lethal infection with a similar magnitude as that achieved by mammalian cell-produced E16 (plant-E16, IC$_{50}$ = 0.19 µg, mammalian-E16, IC$_{50}$ = 0.15 µg, P >0.6) (Fig. 4.2).

For post-exposure therapeutic studies, mice were passively administered a single dose (4 to 100 µg) of E16 2 or 4 days after subcutaneous (s.c.) inoculation of 10^2 PFU of WNV (Fig. 4.2c, d). The results indicate that 20 µg of plant-E16 administered 2 days after WNV inoculation protected mice from lethal infection, and a single dose of as low as 4 µg also prevented mortality. Moreover, a single administration of 50 µg of plant-E16 protected up to 70% of mice from lethal infection and a 90% survival rate was achieved with a single 500 µg dose even after 4 days WNV inoculation. This is highly significant since WNV spreads to the CNS in mice by day 4 after infection. This result indicates that plant-derived E16 is efficacious even after WNV entered the brain. Therefore, plant-produced E16 appeared to be a potent post-exposure therapeutic against lethal WNV infection in mice. This is the

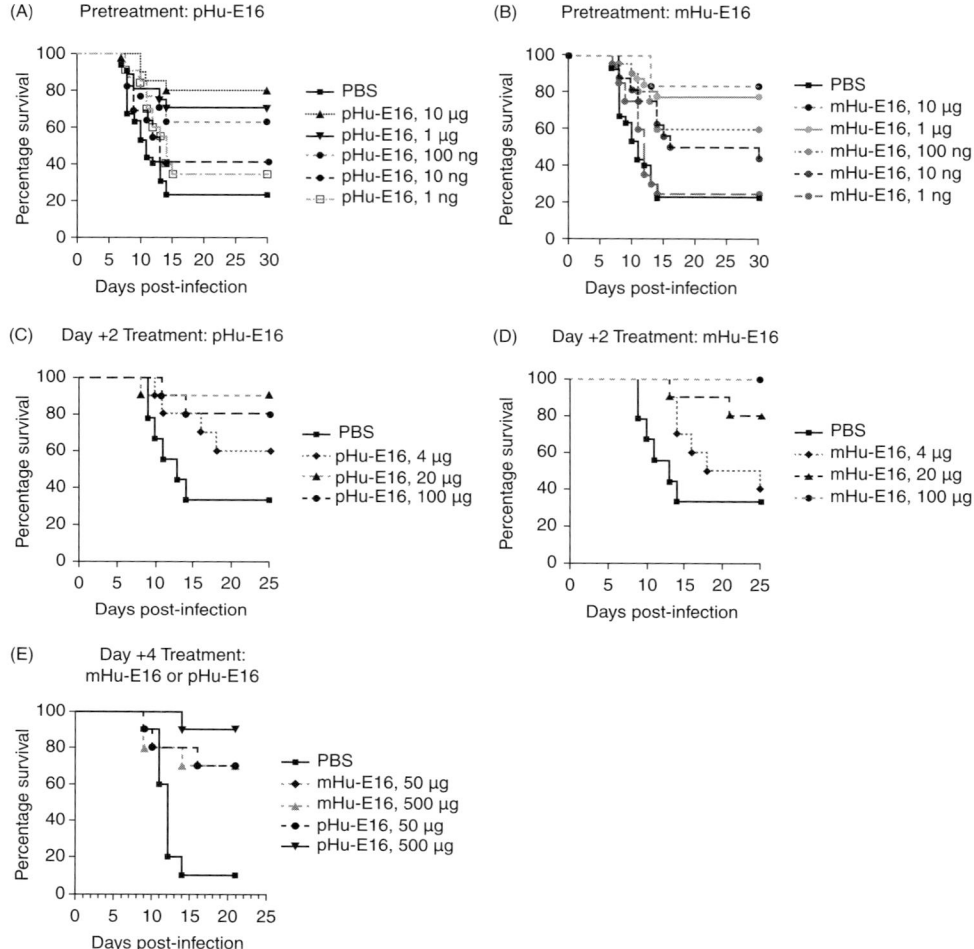

Fig. 4.2. Plant-derived E16 (pHu-E16) protected mice from lethal challenges of WNV equivalently as E16 produced from mammalian cells (mHu-E16) (Lai *et al.*, 2010). (A–B) Mice were passively transferred saline or serial tenfold increases in dose of pHu-E16 (A) or mHu-E16 (B) via intraperitoneal route on the same day as subcutaneous infection with 10^2 pfu of WNV. (C–E) Mice were infected with 10^2 pfu of WNV and then given a single dose of the indicated doses of pHu-E16 or mHu-E16 via i.p. route at (C and D) day +2 or (E) day +4 after infection. Survival data from at least two independent experiments (*n* = 20 per dose) were analysed by the log-rank test.

first result to demonstrate the efficacy of a plant-produced MAb against a lethal infection several days after exposure. Recently, novel variants of E16 including scFv-E16 and tetravalent E16 have been produced in plants (He *et al.*, 2014a, b; Lai *et al.*, 2014). These E16 have shown enhanced neutralization and efficacy, as well as improved feasibility for large-scale manufacturing (He *et al.*, 2014b; Lai *et al.*, 2014).

In addition to its tremendous potential as a therapeutic agent, the high specificity of recognition between E16 and WNV DIII also makes it a superb candidate for WNV diagnostics. Studies by our research group demonstrated that plant-derived E16 is an effective detection and diagnostic agent for WNV and its infection (He *et al.*, 2012). In contrast to developing therapeutic targets in which the 'KDEL' tetrapeptide ER-retention signal was added to the C-terminus of the HC (Lai *et al.*, 2010) or E16 was expressed in 'humanized' transgenic plant lines (Lai *et al.*, 2014), diagnostic reagents are used *in vitro* and the concern for the potential adverse host immune response becomes irrelevant. Thus, E16 was produced with a HC lacking the extra KDEL peptide in WT *N. benthamiana* and lettuce plants. The ease of using WT plants to produce E16 greatly simplifies the experimental procedure. More importantly, it minimizes biosafety and regulatory concerns and costs associated with transgenic crops and GM plants (Chen, 2011b). One critical issue of immuno-based WNV diagnosis is the cross-reactivity of antibodies among flaviviruses. The high specificity of plant-derived E16s may offer a tool to resolve this problem. For example, plant E16 can be used in a VecTest-like WNV antigen assay to rapidly detect WNV infection in wild bird and mosquito populations (Panella *et al.*, 2005). Unlike the current kit, which cannot distinguish WNV from other flaviviruses, using E16 in this assay will reduce ambiguity and improve its accuracy. E16 can also be used in assays based on epitope-blocking ELISAs, which is believed to be a promising method for both disease diagnosis and surveillance. We speculate that the high specificity of E16 to WNV DIII will enhance the accuracy of such assays. In addition, using plant-derived E16 for this method will address the issue of production scalability and MAb production cost.

It is worth noting the use of lettuce for E16 production (Lai *et al.*, 2012). Even though tobacco and related *N. benthamiana* are the most common host plants for expressing biologic proteins, most tobacco and other *Nicotiana* plants contain high levels of phenolics and toxic alkaloids, which foul purification resins and are difficult to remove in downstream processing (Roque *et al.*, 2004; Platis and Labrou, 2008). The biomass yield and production speed of lettuce are just as robust as tobacco plant and it is already cultivated commercially in very large scales. As demonstrated in our studies, the purification procedure of E16 from lettuce is greatly simplified due to low quantities of phenolics and alkaloids in lettuce leaves (Lai *et al.*, 2012). As a result, the life cycle of protein A resin is prolonged and the overall production cost of MAbs is reduced. The agricultural and food industries have already established the infrastructure and technology for large-scale lettuce growing and processing. Such infrastructure and technology can be rapidly adapted for MAb production, providing another advantage for using lettuce as a production host. In addition to being the first demonstration of functional therapeutic and diagnostic activities for a lettuce-produced MAb, our studies also serve as a proof-of-principle for using commercially produced lettuce to produce biologics. Our success in producing high-levels of E16 with commercially produced lettuce and geminiviral vectors suggests that biomass production could be subcontracted to existing commercial growers to provide inexpensive lettuce with consistent growth and postharvest storage conditions to further increase the MAb production yield. This will eliminate the need for capital investment in purpose-built biomass facilities, but allows access to potentially unlimited quantities of inexpensive plant material for large-scale production.

Overall, studies of E16 by our group demonstrate that plant-derived MAbs can function effectively as a post-exposure therapy against a potentially lethal infectious disease. The rapid high-level production and assembly of E16 by deconstructed viral vectors, the development of a scalable and cGMP compliant downstream processing scheme, and the access to unlimited quantities of inexpensive commercial plant material convincingly demonstrate the viability of plants as a platform for large-scale and cost-effective production of MAbs. This technology is being applied to develop MAb therapeutics and diagnostics against other flaviviruses such as DV (Q. Chen, manuscript in preparation).

4.4.3 MAbs against filovirus

The filamentous viruses in the family *Filoviridae* belong to two genera: Ebola virus and Marburg virus. Filoviruses cause severe and often fatal haemorrhagic fever in humans and non-human primates. Natural outbreaks of these viruses occur periodically and often cause high rates of local mortality. The risk of a worldwide epidemic has increased as international travel has become more commonplace. The US Centers for Disease Control classify Ebola and Marburg viruses as 'category A' bioterrorism agents that pose a risk to national security. In spite of the potential terrifying consequence of a natural or man-made filoviral epidemic, no licensed therapeutic is currently available.

Ebola virus, the causative agent of Ebola haemorrhagic fever, has a negative sense, single-stranded RNA genome and a filamentous shape approximately 80 nm in diameter (Ellis *et al.*, 1978). Four distinct species of Ebola virus have been identified: Zaire, Sudan, Ivory Coast and Reston, with Zaire Ebola causing up to a 90% mortality rate (Wilson *et al.*, 2001). Currently, there are no countermeasures approved for prevention or treatment of this disease in humans. The urgent need for biodefence and public health readiness has led to a search for therapeutics that could treat this deadly disease. MAbs have been identified as one group of potential therapeutics. Since the use of such MAb therapeutics would be in the service of public health and the creation of a biodefence stockpile, the speed and cost of production are important considerations in their development.

Mouse MAbs to the Ebola glycoprotein (GP) have been generated and several have been shown to protect mice from lethal Ebola virus infection even 2 days after viral exposure (Wilson *et al.*, 2000). Three of the protective MAbs, i.e. 6D8, 13C6 and 12F6, recognizing three distinct and non-overlapping epitopes on Ebola GP, have been investigated as potential candidates for plant-made human therapeutics. First, the murine MAbs were de-immunized by removing potential T-cell epitopes from their variable region sequences through point mutation (Zeitlin *et al.*, 2011). The de-immunized variable regions were then graphed to human

IgG1constant regions. These measures are aimed to reduce the antigenicity of the murine MAb in humans. Our research group has investigated the expression of 6D8 MAb in plants (Huang *et al.*, 2010; Lai *et al.*, 2012). The coding sequence of the LC and HC of 6D8 was optimized for expression in *N. benthamiana* and cloned into geminiviral vectors. In one of the vector configurations, the LC and HC were cloned into two separate vectors. Our ELISA analysis showed that fully assembled 6D8 was detected 2 days post the co-infiltration of LC and HC vectors and reached the peak accumulation at 4 dpi up to 0.5mg g^{-1} LFW (Huang *et al.*, 2010). Our results also demonstrated that there was no proteolytic clipping of the LC and HC, and also, that plant-produced 6D8 was glycosylated. In the second vector configuration, the LC and HC replicons were cloned into a single vector. Infiltration of this multi-replicon vector into *N. benthamiana* resulted in a rapid high-level expression of 6D8 with an accumulation level and kinetics comparable to that produced by the co-infiltration of separate vectors (Huang *et al.*, 2010). Further analysis confirmed that the single vector-produced 6D8 has the correct LC and HC components and is fully assembled. Our laboratory also explored the feasibility of producing 6D8 in commercially produced lettuce with geminiviral vectors (Lai *et al.*, 2012). Our results indicate that 6D8 can be produced in lettuce with a comparable accumulation level and rapidity as in *N. benthamiana* (Lai *et al.*, 2012). Similar to the methods for developing MAb therapeutics against WNV, we also developed a downstream process for extraction and purification of 6D8. From *N. benthamiana* tissue, 6D8 was effectively purified to greater than 90% purity by ammonium sulfate precipitation followed by protein G affinity chromatography (Huang *et al.*, 2010). Because lettuce does not produce traceable amounts of secondary metabolites, the ammonium sulfate precipitation step was eliminated and a single protein A chromatography step purified 6D8 to >95% pure (Lai *et al.*, 2012). The use of lettuce, therefore, simplifies the purification process and lowers the production cost of 6D8. Furthermore, the successful use of commercially produced lettuce demonstrated a viable option of supplying an almost unlimited amount of cost-effective material for large-scale production of 6D8 and other anti-Ebola MAb therapeutics. The functional activity of both tobacco and lettuce-produced 6D8 was examined in a binding assay with irradiated Ebola virus coated on an ELISA plate. Our results demonstrated that plant-derived 6D8 retains its specific affinity for Ebola GP protein, suggesting it is functionally active (Huang *et al.*, 2010; Lai *et al.*, 2012).

De-immunized MAb 13F6 was also explored as a potential immuno-protectant for Ebola virus. 13F6 recognizes the heavily glycosylated mucin-like domain of GP and was shown to provide 100% prophylactic and therapeutic protection in mice (Wilson *et al.*, 2000). As for 6D8, the gene sequence of the LC and HC of 13F6 were codon optimized for plant expression and cloned into the MagnICON vector modules. These 13F6 gene-carrying MagnICON vectors were infiltrated into *N. benthamiana* plants for transient production. The plant-produced 13F6 was purified by a low-PH (4.8) precipitation followed by a protein A and an anion-exchange (Q) membrane-based chromatography (Zeitlin *et al.*, 2011). The efficacy of the plant-produced 13F6 was tested in a well-established lethal Ebola challenge model. Groups of ten mice were injected a single dose (3, 30, or 300 µg) of plant-derived 13F6 through the intraperitoneal (i.p.) route. Mammalian cell-derived 13F6 and PBS buffer were used in parallel as a positive and a negative control, respectively. After 24 h, mice were challenged with a lethal dose (1000 PFU) of Ebola virus by i.p. inoculation. The results of this study showed that overall plant-derived 13F6 was more protective than its mammalian counterpart (ED_{50} = 3 µg versus ED_{50} = 11 µg) (Zeitlin *et al.*, 2011). For example, in the low MAb-dose groups (3 µg, ~0.2 mg kg^{-1}), mice injected with plant-13F6 were significantly more protected (median survival = 18.5 days) than those that received mammalian-13F6 (P <0.05; median survival = 7 days) or PBS buffer control (P <0.05; median survival = 6 days) (Zeitlin *et al.*, 2011). The significant improvement of the potency is due to the fact that this particular plant-derived 13F6 has a more uniform glycosylation pattern, in contrast to the heterogeneous glycan populations of the mammalian-13F6 (see Glycosylation section below).

The efficacy of plant-derived anti-GP MAbs was further investigated in non-human primates by using a rhesus macaque model due to its resemblance to the human disease in both symptom onset and progression (Bente *et al.*, 2009). In this study, 6D8, 13C6 and 13F6 were produced in *N. benthamiana* by using MagnICON vectors (Olinger *et al.*, 2012). Each MAb was purified with a regime of protein A affinity-anion exchange Q membrane chromatography and polished by either a CHT Ceramic Hydroxyapatite Type I (for 13C6 and 6D8) or a MEP Hypercel (for 13F6) chromatography (Olinger *et al.*, 2012). The three MAbs were then combined in the 1:1:1 ratio to form a treatment mixture termed 'MB-003' (Olinger *et al.*, 2012). In the first animal experiment, rhesus macaques were challenged intramuscularly (i.m.) with 100 PFU Ebola virus (Zaire Kikwit strain) and treated initially 1 h post-infection (p.i.) with mammalian cell- or plant-produced MB-003 at a dosage of 50 mg/kg/MB-003, and subsequently with two additional doses on days 4 and 8 p.i., respectively. The result of this first pilot study was encouraging: no symptoms were observed and no virus was detected in all MB-003 treated animals. In contrast, macaques treated with buffer or a negative control MAb showed symptoms of infection and eventually died (Olinger *et al.*, 2012). Based on this result, a higher dose of challenge (1000 PFU) was used in the second study. The MAb cocktail treatment regimen was the same as in the first study and the treatment mixture used was either mammalian cell-produced MB-003 (50 mg kg^{-1} MB-003) or a significantly lower dose of plant-derived MB-003 (16.7 mg kg^{-1} MB-003). All three animals treated with the lower dose of plant-MB-003 survived the challenge and showed no detectable level of virus. Surprisingly, only one of the two mammalian-MB-003 treated macaques survived in spite of the three times higher MAb dosage (Olinger *et al.*, 2012). This study corroborated the results from the earlier mice studies that plant-derived MAbs have a superior potency to their mammalian-produced counterparts

(Olinger *et al*., 2012). The successful protection by the plant-derived MB-003 led to the third experiment, in which the initiation of treatment was delayed to 24 h or 48 h after the viral challenge, and then followed by three additional doses of plant-derived MB-003 (16.7 mg kg^{-1} MB-003) on days 5, 8 and 10 p.i. (for the 24-h p.i. group) or 6, 8 and 10 p.i. (for the 48-h p.i. group). Two out of two control macaques, which were treated with either an irrelevant MAb or buffer, had high viral titres and showed significant drops in platelet and glucose levels, and died day 7 p.i. (Olinger *et al*., 2012). In contrast, two out of the three plant-MB-003 treated animals in both the 24-h p.i. and the 48-h p.i. groups survived and they displayed no signs of infection. Furthermore, these surviving macaques had very little change in blood chemistry and their serum viral titres were 100,000-fold less than that of controls (Olinger *et al*., 2012). Therefore, these studies demonstrated the efficacy of the plant-derived MAb cocktail in protecting macaques from lethal Ebola infection even when administered 24 or 48 h after viral challenge. This represents successful post-exposure *in vivo* efficacy in a non-human primate by a plant-produced MAb mixture and suggests that MAbs should be further explored as a potential therapeutic for Ebola infection.

4.4.4 MAb-based vaccine for Ebola virus

Despite the potential threat of Ebola epidemics, there is no licensed vaccine currently available to prevent this detrimental disease in humans. Current vaccine candidates include Ebola Venezuelan equine encephalitis virus replicon particles (VRP) (Pushko *et al*., 2000), a recombinant vesicular stomatitis virus-based Ebola vaccine (Geisbert *et al*., 2009) and an adenovirus-based vaccine (Richardson *et al*., 2009). These vaccine candidates have demonstrated their protective potency in non-human primates against lethal Ebola challenge. However, they are based on genetically modified live viruses, which have a tendency to lose potency over a period of time. Furthermore, they demand stringent storage conditions, which may pose challenges in their practical application. For example, vaccines in biodefence stockpiles are required to have long-term stability during storage and distribution with minimal cold chain requirements. As a result, these vaccines are considered poor candidates for such a purpose. Thus, the need for national security and public safety demands the search for vaccines that are not only effective and stable, but can also be produced in a production platform that is rapid and cost effective.

A previous study has demonstrated that plant-derived recombinant immune complexes can be a strong vaccine candidate for infectious diseases (Chargelegue *et al*., 2005). In the study, the coding sequence of tetanus toxin fragment C (TTFC) was fused in frame to that of the C-terminus of the HC of an anti-TTFC IgG. When co-expressed with its LC in tobacco, this HC-fusion polypeptide assembled into an IgG-like structure with two antigen (TTFC)-binding sites at its antigen binding (Fab) domains, and their cognate antigen (TTFC) on the opposite end of the molecule. When the Fab domains of the molecule bind to TTFC on other molecules, they form a repeated array called recombinant immune complex (Fig. 4.3). Like a native immune complex formed between an antigen and an antibody, the recombinant TTFC immune complexes were shown to have enhanced binding to C1q and Fc receptors than TTFC alone (Chargelegue *et al.*, 2005). This enhancement should facilitate their pick up by antigen presenting cells and promote the efficient presentation of antigen peptides on major histo-compatibility complex class I and II (Regnault *et al.*, 1999). The enhanced endocytosis, the improved antigen presentation and, thereby, enhanced T cell activation, suggest the potential of immune complexes as effective self-adjuvanting vaccines. Indeed, the recombinant TTFC immune complexes induced a much stronger TTFC-specific immune response than TTFC in a mouse model (Chargelegue *et al.*, 2005). This success has led our laboratory and collaborators to pursue a similar strategy for Ebola vaccine development.

The Ebola recombinant immune complex (EIC) we designed is based on the extracellular subunit of GP (GP1) and 6D8 that specifically recognizes a linear epitope on GP1. To produce a polypeptide with GP1 fused to the C-terminus of 6D8 MAb, the coding sequence of GP1 was first optimized *in silico* with tobacco-preferred codons and then fused to the 3' end of the sequence coding for 6D8 HC. This HC-GP1 fusion construct was then cloned into geminiviral expression vectors along with the 6D8 LC. We chose to express the EIC with the geminiviral vector over the MagnICON vectors, because we were uncertain that the latter would provide the appropriate

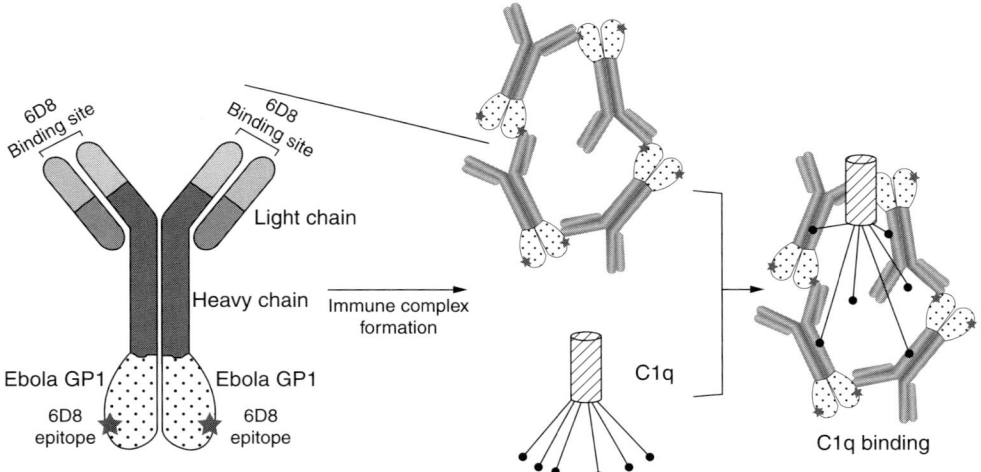

Fig. 4.3. Diagram illustrating the potential structure of the recombinant IgG-GP1 fusion and its assembly to form Ebola immune complexes (EIC). (Left) The 6D8 HC-GP1 fusion protein is assembled with the LC to form a chimeric IgG-GP1. Medium grey, HC chain; dotted, GP1; black star, 6D8 epitope. (Centre) The epitope-binding sites at the top of the molecule bind to the 6D8 epitope on other chimeric molecules, which results in EIC complex assembly. Dark grey, IgG components. (Right) The complement component C1q binds to the Fc region of IgG molecules that are bound to antigen. (Phoolcharoen *et al.*, 2011a, with permission from John Wiley and Sons.)

pairing for heterodimeric molecules like the EIC with two monomers of drastic size difference. Two versions of the geminiviral vectors were used for expressing the EIC: the LC and HC-GP1 fusion replicons in two separate vectors or in a dual-replicon single vector. Transient expression with these vectors in *N. benthamiana* indicated that the dual replicon vector drove a higher expression level (~50 µg IgG equivalent/LFW) of the IgG-GP1 fusion molecule than co-delivery of the two separate replicons and the highest level of accumulation occurred at 4 dpi (Phoolcharoen *et al.*, 2011a). Our analysis showed that the HC-GP1 fusion construct produced a 130 kDa band on SDS-PAGE gels under reducing conditions (Phoolcharoen *et al.*, 2011a). This is the predicted size for the HC-GP1 fusion polypeptide. As expected, the LC was detected as a 25 kDa band. Under non-reducing conditions, a single high molecular weight band (>250 kDa) was detected with 13C6 MAb. This result indicated that the HC-GP1 and LC polypeptide correctly assembled into a tetrameric IgG-like structure. Since 13C6 recognizes a conformational epitope of GP1, this result also demonstrated that GP1 folded properly to form authentic antigen in the fusion protein.

The 6D8 IgG-GP1 fusion protein was purified by ammonium sulfate precipitation and protein G affinity chromatography. As in the case with plant-produced MAbs, this two-step process enriched the fusion protein to ~90% pure (Phoolcharoen *et al.*, 2011a). However, multiple degradation products were observed in the purified samples, suggesting proteolytic clipping during extraction and purification. This issue is being addressed by a more rigorous downstream processing development study (Q. Chen, manuscript in preparation). As the assembled 6D8 IgG-GP1 fusion protein has two ends that have high binding affinity to each other, immune complexes can be formed when the 6D8 antigen binding sites at the N-termini encounter the GP1 6D8 epitope at the C-termini of another fusion molecule (Fig. 4.3). Multiple fusion proteins can potentially link together like this to form a large network of recombinant immune complex (Fig. 4.3). Analyses with dynamic light scattering confirmed the EIC formation for the 6D8 IgG-GP1 fusion protein, and suggested that each complex contains four fusion proteins (Phoolcharoen *et al.*, 2011a). Result of size-exclusion chromatographic analysis corroborated this interpretation. The formation of the EIC was further supported by functional analysis with C1q binding assay (Fig. 4.3). It was shown that the plant-derived EIC sample had more enhanced binding to C1q than that of the parent 6D8 MAb (Phoolcharoen *et al.*, 2011a). The immunogenicity of EIC was examined in a mouse model. Four doses of EIC (10 µg) were injected s.c. on days 0, 21, 42 and 63 along with VRP-GP1 as the positive control and PBS buffer as the negative control. The results showed that EIC evoked Ebola-specific antibody response and the antibody titres were comparable to, but somewhat less than, those of the VRP-GP1 positive control (Phoolcharoen *et al.*, 2011a). This was an encouraging result, as VRP-GP1 showed complete protection of mice and guinea pigs from Ebola challenge (Pushko *et al.*, 2000). We speculated that the difference in adjuvant may be the cause of the titre variation between the two vaccine candidates. VRP-GP1 is an alphavirus vector and undergoes limited replication in antigen-presenting cells of the immunized host, thus creating viral replicon RNA – a powerful adjuvant through toll-like receptor (TLR) signalling (Alexopoulou *et al.*, 2001). In contrast, no adjuvant was used for EIC in this study. This interpretation was supported by the result of a later study that inclusion of adjuvants, e.g. a TLR agonist polyinosinic:polycytidylic acid (PIC), significantly boosted the titres of anti-EIC antibodies in mice (Phoolcharoen *et al.*, 2011b).

The efficacy of the plant-derived EIC was further examined in a mouse model. Mice were vaccinated with the same regimen as discussed above except PIC (20 mg) or PIC (20 µg) + alum were used as adjuvants. Mice were challenged with a lethal dose (1000 PFU) of mouse-adapted Zaire Ebola virus 21 days after the final vaccine dose. Our results indicated that the inclusion of adjuvant PIC significantly boosted the Ebola-specific antibody response as mice vaccinated with EIC + PIC or EIC + PIC + alum showed significantly higher anti-Ebola IgG titre than EIC alone (Phoolcharoen *et al.*, 2011b). In fact, the levels of neutralizing antibody titre from these two vaccination regimens are higher than that of the GP-VRP positive control. Further isotype analysis showed that EIC co-delivered with PIC induced a mixed Th1/Th2 response, while EIC alone triggered mostly a Th2 response (Phoolcharoen *et al.*, 2011b). During the 14 day post-challenge period, weight loss was observed in mice for all of the vaccination groups, indicating their lack of sterilizing immunity. As for the buffer negative control group, none of the mice vaccinated with EIC alone survived the viral challenge. However, co-delivery of EIC and PIC or EIC, PIC and alum protected 80% of mice against lethal challenge, similar to that of the GP-VRP positive control group (Phoolcharoen *et al.*, 2011b). These results indicate that levels of IgG and neutralizing antibody specific to Ebola correlates with protection. Moreover, the results also suggest that protection also corresponds with the IgG subtype profile, particularly with the amount of IgG2a. These studies, therefore, demonstrated the potential of EIC as a vaccine strategy for filoviruses. Furthermore, the plant production of EIC also overcame the longstanding technical hurdle of producing GP1, as it is highly toxic to the production host cell. Our fusion strategy not only resolved the issue of GP1 production, but also provides an IgG scaffold that allows its easy purification. As a result, plant-derived EIC also enhances filovirus vaccine manufacturing, making it highly relevant to a national biothreat reduction stockpile.

4.4.5 Idiotype antibody-based vaccine for non-Hodgkin's lymphoma

Non-Hodgkin's lymphoma (NHL) is a group of blood cancers that have significant impact on global health. For example, NHL resulted in 210,000 deaths in 2010 around the world, and the number of new cases for 2012 was estimated to be over 60,000 in the USA alone. In NHL, each malignant B cell clone expresses a unique cell surface immunoglobulin (Ig) as the tumour-specific marker. As a result, standard treatments are not effective for NHL patients, and many of them suffer a relapse. The extremely variable nature of NHL calls for the development of patient-specific cancer treatments. As such, a recent strategy focuses on active immunotherapy through vaccination with the patient's own idiotype Ig. It was shown that when conjugated to a strong immunogen such as keyhole limpet haemocyanin (KLH) and co-administered with granulocyte-macrophage colony stimulating factor (GM-CSF) as adjuvant, idiotype Ig induced an antigen-specific immune response in a large percentage of patients. Furthermore, a superior clinical outcome in remission and survival has been observed in the responders to Ig vaccination (Bendandi, 2006).

The development of individualized vaccines requires a production platform that offers short production time, low-cost products and the flexibility to rapidly produce tailor-made, patient- and tumour-specific antigens. These requirements present a difficult obstacle for mammalian cell culture-based production platforms, but can be readily met by plant production platforms based on transient expression. In fact, an idiotype-specific scFv of the Ig from the 38C13 mouse B cell lymphoma was expressed in *N. benthamiana* plants with a TMV-derived vector. It was shown that 38C13 scFv was produced at high levels in plant leaves within 2 weeks of vector inoculation (McCormick *et al.*, 1999). Analysis also indicated that the scFv folded correctly into a conformation that is equivalent to the native IgM molecule (McCormick *et al.*, 1999). The scFv was readily recovered in the extracellular fluid and purified to high purity by affinity chromatography. Mice were vaccinated with three doses (15 µg per dose) of scFv (alone or with 10 µg QS-21 adjuvant) via the s.c. route at 2-week intervals. Two doses of KLH conjugated-38C13 IgM (50 µg per dose) were administered s.c. with QS-21 as a positive control. Two weeks after the last vaccination, mice were challenged with a lethal dose of the syngeneic 38C13 tumour (200 cells) by i.p. injection. It was shown that the scFv co-delivered with QS-21 induced a strong anti-38C13 IgG1 response after the third dose (with an average titre of 105 µg ml^{-1}), comparable to that of the positive control with 38C13 IgM-KLH (116 µg ml^{-1}). In addition, 38C13 scFv + QS-21 also triggered an IgG2a isotype response with similar amplitude as the positive control, which has been shown to correlate with augmented tumour protection. Indeed, 90% of mice vaccinated by scFv with QS-21 were protected against the lethal challenge, similar to that of the 38C13 IgM-KLH + QS-21 'gold standard' (McCormick *et al.*, 1999). In contrast, mice that received the negative control vaccination (QS-21 alone) developed abdominal tumours within 15 days of challenge and died within 21 days. Interestingly, mice receiving 38C12 scFv without the adjuvant were also protected equally as the positive control despite the lower levels of antibody titres and the lack of IgG2a response (McCormick *et al.*, 1999). This study demonstrated the rapidness of the plant production system in generating tumour-specific protein vaccines, suggesting that it is a viable strategy for producing treatment for NHL. In a follow-up study, 44 human tumour-specific scFvs were rapidly expressed in *N. benthamiana* plants, 38 of which showed high-levels of expression (McCormick *et al.*, 2003). This presents an 86% production rate, which is comparable to that of standard hybridoma methods. Like the murine 38C13 scFv, the secreted human scFv proteins were also shown to fold into a conformation that is relevant to the tumour Ig by three independent immunological tests. Eight of these scFvs were tested in a mouse model and were shown to induce appropriate idiotype responses with minimal cross-reactivity to irrelevant Igs or scFvs (McCormick *et al.*, 2003). Furthermore, anti-scFv sera were also shown to recognize the Ig on human tumour cells (McCormick *et al.*, 2003). This study not only further demonstrated the speed of the plant expression system, but also showcased the flexibility and versatility of the system in producing multiple personalized human vaccines.

The safety and immunogenicity of these scFv-based personalized vaccines were tested in a phase I clinical trial (McCormick *et al.*, 2008). Sixteen patients who had follicular B-cell lymphoma completed this trial. Their tumour-specific Ig HC and LC genes were cloned and the resulting scFv were successfully expressed in *N. benthamiana*. These scFvs were extracted and purified under FDA cGMP guidelines, and their identity, tumour relevance, purity and potency were certified to meet the required specifications by the quality management system (QMS) (McCormick *et al.*, 2008). Vaccination was initiated 6 months after the last cycle of chemotherapy and given by s.c. once a month for a total of six doses. Patients were divided into four cohorts of four patients each: cohort A received 0.2 mg scFv, B received 2.0 mg scFv, C received 0.2 mg scFv + GM-CSF adjuvant, and D received 2.0 mg scFv + GM-CSF. The safety results indicated that all 16 individualized scFv vaccines were well tolerated by all patients. There were no serious adverse events observed during the study (McCormick *et al.*, 2008). As a result, the study was completed by all patients due to the lack of adverse events caused by the vaccine. Greater than 70% of the patients developed humoral or cellular immune responses and 47% of them are antigen-specific (McCormick *et al.*, 2008). In one patient, a specific cellular response was detected after the second vaccine dose and was found to sustain at high levels up to 7 months after the last vaccination (McCormick *et al.*, 2008). Thirteen of 16 patients remained alive with a median follow-up of 78 months. Collectively, these findings suggest that plant-produced idiotype vaccines are feasible to produce, safe to administer, and provide a viable option for idiotype-specific immune therapy in follicular lymphoma patients.

To further optimize the up- and downstream manufacturing process for these vaccines, new plant expression systems and a whole Ig strategy were explored. In one of these studies, the LC and HC variable region of a patient's tumour-specific Ig was fused genetically to the constant region of kappa or lambda LC and IgG1 HC, respectively. Twenty fusion tumour-specific antibodies like this were then produced with the MagnICON vectors in *N. benthamiana* (Bendandi *et al.*, 2010). The results indicated that using MagnICON vectors for transient expression in plants drastically improved the yield of these vaccine candidates, up to 1000-fold higher than the previous expression method (Bendandi *et al.*, 2010). The inclusion of the Fc region of a MAb in the full antibody construct also allowed convenient purification of the vaccine candidates with consistency in purity and recovery. Overall, the manufacturing process became reliable and robust, requiring only 2 weeks for vaccine expression and purification and less than 12 weeks from biopsy to vaccination (Bendandi *et al.*, 2010). A phase I human clinical trial has been initiated to test safety and immunogenicity of the MagnICON-expressed full idiotype MAb antigen candidates. Twelve patients were enrolled and subjected to 12 doses of KLH-conjugated Ig (0.5 mg Ig + 0.5 mg KLH per dose) monthly for the first eight doses and bimonthly for the last four doses. While this trial is still in progress, earlier reports indicated that the full idiotype MAb-based vaccine was well-tolerated in patients (Whaley *et al.*, 2012).

4.5 Glycosylation of Plant-made MAbs

Glycosylation of MAbs has a major impact on the pharmacokinetics, antigen binding, stability, effector functions and efficacy of the MAbs. Glycosylation is dependent on a series of post-translational modification steps by host cells. As such, MAbs produced by heterologous systems may have appreciable structural and functional differences from the native molecules and could be immunogenic in humans. Thus, the glycosylation status of MAbs is critical for their structure and function, as well as determines the choice of platforms for their production. For example, currently, full MAbs cannot be produced in *Escherichia coli* cultures. As prokaryotic cells, *E. coli* do not have the required endomembrane system for the proper glycosylation and assembly of MAbs. Cultures based on insect cells produce MAbs in predominantly high mannose glycoforms, which is not an optimal system for MAb production (Harrison and Jarvis, 2006). Even in mammalian cells, the glycosylation profile of MAbs is highly heterogeneous. The presence of mixed and inconsistent glycosylation types is a common phenomenon for most mammalian cell-produced glycoproteins (Hossler *et al.*, 2009). Thus, none of the current popular production systems is ideal for the production of MAbs.

N-glycosylation of proteins in plants is generally similar to that of mammalian cells. However, WT plants add plant-specific β-1,2-xylose and core α-1,3-fucose residues to complex N-linked glycans and lack terminal β1,4-Gal and N-acetylneuraminic acid (Neu5Ac) residues (Gomord *et al.*, 2010). These minor differences in protein glycosylation between WT plant and mammalian cells have become one of the major issues of plant-expression platforms, because they may produce improper glycoforms reducing efficacy. But there is also the possibility of inducing plant-glycan specific immune responses in humans that can accelerate protein clearance from plasma or cause potential adverse effects through immune complex formation. Studies have shown that variations in glycosylation patterns do not always lead to loss of *in vitro* or *in vivo* function, or cause side-effects in humans. For example, a tobacco-produced mouse MAb did not induce detectable levels of antibody against its plant-specific glycan in mice (Chargelegue *et al.*, 2000). A similar conclusion was drawn from the phase I clinical trial with plant-produced tumour-specific scFv as vaccines against NHL. Out of the 16 scFv candidates, 13 were found to contain plant-specific glycans. However, no measurable immune responses towards the plant-specific carbohydrates were detected in these patients, but rather their responses were directed to the idiotype sequence itself (McCormick et al., 2008). These studies, therefore, demonstrated that variation in patterns of antigen glycosylation in MAb or its derivatives does not always impair the function or affect the safety of the antibody-based biologics. Despite these observations, a variety of strategies has been developed to circumvent any potential problems associated with the difference between plant and human glycosylation patterns. Glycoengineering is a biological approach to 'humanize' the glycosylation pathway in plants by genetically suppressing or eliminating enzymes for making plant-specific glycans and by introducing glycoenzymes from mammalian cells. Glycoengineering efforts have led to the creation of a portfolio of 'humanized' plant lines that produce MAbs that not only lack plant-specific glycans but with glycoforms that are essentially mammalian (Bosch et al., 2013). In addition, plant glycoengineering also offers an option of creating MAbs with tailor-made glycoforms based on their specific functional needs. These include various specific and defined mammalian glycoforms such as high mannose, GnGn, G0-G2 galactose, bisected GlcNAc, fucosylated and nonfucosylated, and full complex forms with terminal sialic acids (Bosch et al., 2013). In the last few years, the feasibility of using glycoengineered plant lines at biomanufacturing scales was also demonstrated. For example, a double knockout (ΔXF) N. benthamiana plant line, which was created by RNAi to suppress the expression of α-1,3-fucosyltransferase and β-1,2-xylosyltransferase, was used in manufacturing runs of 500 kg scale leaf material to produce anti-Ebola 13C6, 13F6 and 6D8 MAbs (Strasser et al., 2008; Zeitlin et al., 2011). It was shown that MAbs produced in the ΔXF plant line have no plant-specific N-glycans. In fact, 90% of plant-derived MAb had the predicted mammalian glycoform GnGn (Zeitlin et al., 2011). In contrast, mammalian cell-produced MAbs had a mixture of from three to five glycoforms with the most dominant glycoform (G0) ranging from 35 to 53% (Zeitlin et al., 2011). Furthermore, plant-produced 13F6 has higher affinity to the Fc receptor of FcγRIII than its mammalian cell-produced counterpart. Correspondingly, the plant-derived 13F6 showed increased protection by a factor greater than twofold more than the mammalian cell-produced 13F6 (Zeitlin et al., 2011). The significant improvement of plant-13F6's potency can be attributed to its highly uniform glycoform of GnGn, which lacks the core fucose. In contrast, mammalian cell-produced 13F6 had a mixture of heterogeneous glycan populations, 82% of which have the core fucose (Zeitlin et al., 2011). This result is in line with a previous observation that elimination of core fucose dramatically improves the antibody-dependent cellular cytotoxicity activity of MAbs mediated by natural killer cells (Umana et al., 1999). The superior potency of plant-derived Ebola protective MAbs to their mammalian-produced counterparts was further demonstrated in a later challenge study with non-human primates. The results showed that plant-produced MB-003, which contained 13C6, 13F6 and 6D8 MAbs with a highly homogeneous (>80%) GnGn glycoform, were more protective against a lethal Ebola challenge than the mammalian cell-produced MB-003, even at one-third of the dose of the latter (Olinger *et al.*, 2012).

For certain classes of MAbs, the presence of the terminal galactose or sialic acid residues is required for their biological activities and stability. Currently, only very limited control of this important modification is offered by manufacturing technologies based on mammalian cell culture (Hossler *et al.*, 2009). Again, glycoengineered plants have provided a solution to overcome this difficult challenge and are being used to manufacture galactosylated and sialylated HIV MAbs (Strasser *et al.*, 2009; Castilho *et al.*, 2010). Two plant-derived anti-HIV MAbs (2G12 and 4E10) were shown to be fully galactosylated and exhibited much improved viral neutralization potency compared with those produced in mammalian cells (Strasser *et al.*, 2009). This advantage of the plant expression systems can be applied to produce superior biologics beyond the realm of MAbs. For example, glucocerebrosidase (GCD) is a therapeutic enzyme for treating Gaucher's disease, and the very first plant-derived biologic ever approved by FDA (1 May 2012). Mammalian cell-produced GCD requires *in vitro* N-glycan processing to achieve the desired efficacy, while the carrot cell-produced GCD (commercial name: ELELYSO™) already has the required glycoform, eliminating the costly N-glycan processing, and possibly resulting in better and more consistent efficacy (Faye and Gomord, 2010).

Overall, these studies have demonstrated two additional advantages of the plant expression system. First, MAbs produced from glycoengineered plants possess a high degree of glycan uniformity that cannot be matched by mammalian cells or achieved by *in vitro* treatments. This is highly significant, as the ability to obtain a consistent glycoform profile in production is desirable from both a quality and regulatory perspective. Second, this plant production platform also offers opportunities to create biobetters with superior efficacy and safety than those produced by the traditional platforms. At least in theory with glycoengineered plant lines, one can custom design MAbs with a tailor-made glycoform that is best suited for its efficacy or safety depending on its particular clinical application. Overall, the creation of these glycoengineered plants not only eliminates concerns for potential undesirable effects of plant-specific glycans, but also provides additional advantages for plant-based expression systems.

4.6 Conclusions

Enormous progress has been made in recent years in the field of plant-made MAbs. The development of deconstructed viral vectors has rejuvenated this field and provided a plant-production platform with high MAb yield, rapid production speed, high scalability, low cost, and high flexibility and versatility. As a result, a wide spectrum of MAbs has been successfully produced in a variety of plants. Many of which have shown the proper assembly, effective *in vitro* neutralization and potent *in vivo* efficacy in animal models. Many plant-derived MAbs have been produced in large manufacturing scales under the FDA cGMP regulations, and shown to meet FDA quality standards in identity, purity and potency. Recent strides in plant glycoengineering provide additional advantages far beyond the traditional benefits of low cost, high scalability and increased safety, which limit the application to producing biosimilars. Because of this breakthrough, plants provide an excellent platform in producing MAbs with a defined mammalian glycoform, and are being explored to develop MAb biobetters with far more superior efficacy and safety profiles. Several such plant-produced MAb or MAb derivatives have or are being examined in human clinical trials. A new era for plant-made MAbs has been marked by the recent FDA approval of a carrot cell-derived GCD, silencing the lingering criticism of plant-based expression platform due to the absence of FDA-approved human products. We speculate that plant expression strategies based on deconstructed viral vectors hold the greatest potential for producing affordable, high quality MAbs to meet the ever-increasing global demand. They will offer superior capacity, safety, time- and cost-saving benefits for biosimilar MAb manufacturing, and enormous opportunities for biobetter MAb development.

Acknowledgements

The authors wish to thank Dr J. He, W. Phoolcharoen, M. Hansen, P. Li, J. Hurtado, J. Stahnke, K. Leuzinger, R. Sun and other members of the Chen laboratory for their contributions to the projects described. The critical reading of the manuscript by J. Caspermeyer is appreciated. Projects performed by our laboratory were supported in part by NIH-NIAID grants U01 AI075549 and 1R21AI101329 to Q. Chen.

References

Alexopoulou, L., Holt, A.C., Medzhitov, R. and Flavell, R.A. (2001) Recognition of double-stranded RNA and activation of NF-kappaB by Toll-like receptor 3. *Nature* 413(6857), 732–738.
Barta, A., Sommergruber, K., Thompson, D., Hartmuth, K., Matzke, M.A. and Matzke, A.J.M. (1986) The expression of a nopaline synthase - human growth hormone chimaeric gene in transformed tobacco and sunflower callus tissue. *Plant Molecular Biology* 6(5), 347–357.
Bendandi, M. (2006) Clinical benefit of idiotype vaccines: too many trials for a clever demonstration? *Reviews of Recent Clinical Trials* 1(1), 67–74.
Bendandi, M., Marillonnet, S., Kandzia, R., Thieme, F., Nickstadt, A., Herz, S., Frode, R., Inoges, S., Lopez-Diaz de Cerio, A., Soria, E., Villanueva, H., Vancanneyt, G., McCormick, A., Tuse, D., Lenz, J., Butler-Ransohoff, J.E., Klimyuk, V. and Gleba, Y. (2010) Rapid, high-yield production in plants of individualized idiotype vaccines for non-Hodgkin's lymphoma. *Annals of Oncology* 21(12), 2420–2427.
Bente, D., Gren, J., Strong, J.E. and Feldmann, H. (2009) Disease modeling for Ebola and Marburg viruses. *Disease Models & Mechanisms* 2(1–2), 12–17.
Bode, A.V., Sejvar, J.J., Pape, W.J., Campbell, G.L. and Marfin, A.A. (2006) West Nile virus disease: a descriptive study of 228 patients hospitalized in a 4-county region of Colorado in 2003. *Clinical Infectious Diseases* 42(9), 1234–1240.

Bosch, D., Castilho, A., Loos, A., Schots, A. and Steinkellner, H. (2013) N-Glycosylation of Plant-produced Recombinant Proteins. *Current Pharmaceutical Design* 19(31), 5503–5512.

Brisson, N., Paszkowski, J., Penswick, J., Bronenborn, B., Potrykus, I. and Hohn, T. (1984) Expression of a bacterial gene in plants by using a viral vector. *Nature* 310, 511–514.

Castilho, A., Strasser, R., Stadlmann, J., Grass, J., Jez, J., Gattinger, P., Kunert, R., Quendler, H., Pabst, M., Leonard, R., Altmann, F. and Steinkellner, H. (2010) In planta protein sialylation through overexpression of the respective mammalian pathway. *Journal of Biological Chemistry* 285(21), 15923–15930.

Chargelegue, D., Vine, N.D., van Dolleweerd, C.J., Drake, P.M. and Ma, J.K. (2000) A murine monoclonal antibody produced in transgenic plants with plant-specific glycans is not immunogenic in mice. *Transgenic Research* 9(3), 187–194.

Chargelegue, D., Drake, P.M.W., Obregon, P., Prada, A., Fairweather, N. and Ma, J.K.C. (2005) Highly immunogenic and protective recombinant vaccine candidate expressed in transgenic plants. *Infection and Immunity* 73(9), 5915–5922.

Chen, Q. (2008) Expression and Purification of Pharmaceutical Proteins in Plants. *Biological Engineering* 1(4), 291–321.

Chen, Q. (2011a) Expression and manufacture of pharmaceutical proteins in genetically engineered horticultural plants. In: Mou, B. and Scorza, R. (eds) *Transgenic Horticultural Crops: Challenges and Opportunities - Essays by Experts*. Taylor & Francis, Boca Raton, Florida, pp. 83–124.

Chen, Q. (2011b) Turning a new leaf. *European Biopharmaceutical Review* 2(56), 64–68.

Chen, Q. and Lai, H. (2013) Plant-derived virus-like particles as vaccines. *Human Vaccines & Immunotherapeutics* 9(1), 26–49.

Chen, Q., Mason, H., Mor, T., Cardineau, G.A., Arntzen, C. and Tacket, C.O. (2009) Subunit vaccines produced using plant biotechnology. In: Levine, M.M. (ed.) *New Generation Vaccines*. Informa Healthcare, New York, pp. 306–315.

Chen, Q., He, J., Phoolcharoen, W. and Mason, H.S. (2011) Geminiviral vectors based on bean yellow dwarf virus for production of vaccine antigens and monoclonal antibodies in plants. *Human Vaccines* 7(3), 331–338.

Chen, Q., Lai, H., Hurtado, J., Stahnke, J., Leuzinger, K. and Dent, M. (2013) Agroinfiltration as an Effective and Scalable Strategy of Gene Delivery for Production of Pharmaceutical Proteins. *Advanced Techniques in Biology and Medicine* 1(1), 9.

Daniell, H. (2006) Production of biopharmaceuticals and vaccines in plants via the chloroplast genome. *Biotechnology Journal* 1(10), 1071–1079.

Diamond, M.S. and Klein, R.S. (2006) A genetic basis for human susceptibility to West Nile virus. *Trends in Microbiology* 14(7), 287–289.

Dietrich, C. and Maiss, E. (2003) Fluorescent labelling reveals spatial separation of potyvirus populations in mixed infected *Nicotiana benthamiana* plants. *Journal of General Virology* 84(10), 2871–2876.

Dugdale, B., Mortimer, C.L., Kato, M., James, T.A., Harding, R.M. and Dale, J.L. (2013) In Plant Activation: An Inducible, Hyperexpression Platform for Recombinant Protein Production in Plants. *The Plant Cell Online* 25(7), 2429–2443.

Duke, J. (1993) Medicinal plants and the pharmaceutical industry. In: Janick, J. and Simon, J. (eds) *New crops*. Wiley, New York, pp. 664–669.

Ellis, D.S., Simpson, I.H., Francis, D.P., Knobloch, J., Bowen, E.T., Lolik, P. and Deng, I.M. (1978) Ultrastructure of Ebola virus particles in human liver. *Journal of Clinical Pathology* 31(3), 201–208.

Farnsworth, N.R., Akerele, O., Bingel, A.S., Soejarto, D.D. and Guo, Z. (1985) Medicinal plants in therapy. *Bulletin of the World Health Organisation* 63(6), 965–981.

Faye, L. and Gomord, V. (2010) Success stories in molecular farming – a brief overview. *Plant Biotechnology Journal* 8(5), 525–528.

Geisbert, T.W., Geisbert, J.B., Leung, A., Daddario-DiCaprio, K.M., Hensley, L.E., Grolla, A. and Feldmann, H. (2009) Single-injection vaccine protects nonhuman primates against infection with marburg virus and three species of ebola virus. *Journal of Virology* 83(14), 7296–7304.

Giritch, A., Marillonnet, S., Engler, C., van Eldik, G., Botterman, J., Klimyuk, V. and Gleba, Y. (2006) Rapid high-yield expression of full-size IgG antibodies in plants coinfected with noncompeting viral vectors. *Proceedings of the National Academy of Sciences USA* 103(40), 14701–14706.

Gomord, V., Fitchette, A.C., Menu-Bouaouiche, L., Saint-Jore-Dupas, C., Plasson, C., Michaud, D. and Faye, L. (2010) Plant-specific glycosylation patterns in the context of therapeutic protein production. *Plant Biotechnology Journal* 8(5), 564–587.

Harrison, R.L. and Jarvis, D.L. (2006) Protein N-Glycosylation in the Baculovirus-Insect Cell Expression System and Engineering of Insect Cells to Produce 'Mammalianized' Recombinant Glycoproteins. *Advances in Virus Research*, Vol. 68. Academic Press, Waltham, Massachusetts, pp. 159–191.

He, J., Lai, H., Brock, C. and Chen, Q. (2012) A novel system for rapid and cost-effective production of detection and diagnostic reagents of West Nile virus in plants. *Journal of Biomedicine and Biotechnology* 2012 (10.1155/2012/106783), 1–10.

He, J., Lai, H., Engle, M., Gorlatov, S., Gruber, C., Steinkellner, H., Diamond, M. and Chen, Q. (2014a) Generation and analysis of novel plant-derived antibody-based therapeutic molecules against West Nile virus. *PLoS ONE* 9(3), e93541.

He, J., Peng, L., Lai, H., Hurtado, J., Stahnke, J. and Chen, Q. (2014b) A plant-produced antigen elicits potent immune responses against West Nile virus in mice. *BioMed Research International* 2014(10.1155/2014/952865), 10.

Hiatt, A. and Pauly, M. (2006) Monoclonal antibodies from plants: a new speed record. *Proceedings of the National Academy of Sciences USA* 103(40), 14645–14646.

Hiatt, A., Cafferkey, R. and Bowdish, K. (1989) Production of antibodies in transgenic plants. *Nature* 342(6245), 76–78.

Hossler, P., Khattak, S.F. and Li, Z.J. (2009) Optimal and consistent protein glycosylation in mammalian cell culture. *Glycobiology* 19(9), 936–949.

Huang, Z., Chen, Q., Hjelm, B., Arntzen, C. and Mason, H. (2009) A DNA replicon system for rapid high-level production of virus-like particles in plants. *Biotechnology and Bioengineering* 103(4), 706–714.

Huang, Z., Phoolcharoen, W., Lai, H., Piensook, K., Cardineau, G., Zeitlin, L., Whaley, K., Arntzen, C.J., Mason, H. and Chen, Q. (2010) High-level rapid production of full-size monoclonal antibodies in plants by a single-vector DNA replicon system. *Biotechnology and Bioengineering* 106(1), 9–17.

Hubalek, Z. and Halouzka, J. (1999) West Nile fever – a reemerging mosquito-borne viral disease in Europe. *Emerging Infectious Diseases* 5(5), 643–650.

Hull, R. and Plaskitt, A. (1970) Electron microscopy on the behavior of two strains of alfalfa mosaic virus in mixed infections. *Virology* 42(3), 773–776.

Kathuria, S., Sriraman, R., Nath, R., Sack, M., Pal, R., Artsaenko, O., Talwar, G.P., Fischer, R. and Finnern, R. (2002) Efficacy of plant-produced recombinant antibodies against HCG. *Human Reproduction* 17(8), 2054–2061.

Klee, H., Horsch, R. and Rogers, S. (1987) Agrobacterium-mediated plant transformation and its further applications to plant biology. *Annual Review of Plant Physiology* 38, 467–486.

Komarova, T.V., Baschieri, S., Donini, M., Marusic, C., Benvenuto, E. and Dorokhov, Y.L. (2010) Transient expression systems for plant-derived biopharmaceuticals. *Expert Review of Vaccines* 9(8), 859–876.

Komarova, T.V., Kosorukov, V.S., Frolova, O.Y., Petrunia, I.V., Skrypnik, K.A., Gleba, Y.Y. and Dorokhov, Y.L. (2011) Plant-made trastuzumab (herceptin) inhibits HER2/Neu+ cell proliferation and retards tumor growth. *PLoS ONE* 6(3), e17541.

Lai, H. and Chen, Q. (2012) Bioprocessing of plant-derived virus-like particles of Norwalk virus capsid protein under current Good Manufacture Practice regulations. *Plant Cell Reports* 31(3), 573–584.

Lai, H., Engle, M., Fuchs, A., Keller, T., Johnson, S., Gorlatov, S., Diamond, M.S. and Chen, Q. (2010) Monoclonal antibody produced in plants efficiently treats West Nile virus infection in mice. *Proceedings of the National Academy of Sciences USA* 107(6), 2419–2424.

Lai, H., He, J., Diamond, M.S. and Chen, Q. (2012) Robust production of virus-like particles and monoclonal antibodies with geminiviral replicon vectors in lettuce. *Plant Biotechnology Journal* 10(1), 95–104.

Lai, H., He, J., Hurtado, J., Stahnke, J., Fuchs, A., Mehlhop, E., Gorlatov, S., Loos, A., Diamond, M.S. and Chen, Q. (2014) Structural and functional characterization of an anti-West Nile virus monoclonal antibody and its single-chain variant produced in glycoengineered plants. *Plant Biotechnology Journal* doi: 10.1111/pbi-12217.

Latham, T. and Galarza, J.M. (2001) Formation of wild-type and chimeric influenza virus-like particles following simultaneous expression of only four structural proteins. *Journal of Virology* 75(13), 6154–6165.

Leuzinger, K., Dent, M., Hurtado, J., Stahnke, J., Lai, H., Zhou, X. and Chen, Q. (2013) Efficient agroinfiltration of plants for high-level transient expression of recombinant proteins. *Journal of Visualized Experiments* (77), doi: 10.3791/50521.

Li, H. and da'Anjou, M. (2009) Pharmacological significance of glycosylation in therapeutic proteins. *Current Opinion in Biotechnology* 20(6), 678–684.

Lico, C., Capuano, F., Renzone, G., Donini, M., Marusic, C., Scaloni, A., Benvenuto, E. and Baschieri, S. (2006) Peptide display on Potato virus X: molecular features of the coat protein-fused peptide affecting cell-to-cell and phloem movement of chimeric virus particles. *Journal of General Virology* 87(10), 3103–3112.

Lico, C., Chen, Q. and Santi, L. (2008) Viral vectors for production of recombinant proteins in plants. *Journal of Cellular Physiology* 216(2), 366–377.

Lim, J.K., Louie, C.Y., Glaser, C., Jean, C., Johnson, B., Johnson, H., McDermott, D.H. and Murphy, P.M. (2008) Genetic deficiency of chemokine receptor CCR5 is a strong risk factor for symptomatic West Nile virus infection: a meta-analysis of 4 cohorts in the US epidemic. *Journal of Infectious Disease* 197(2), 262–265.

Liu, L., van Tonder, T., Pietersen, G., Davies, J.W. and Stanley, J. (1997) Molecular characterization of a subgroup I geminivirus from a legume in South Africa. *Journal of General Virology* 78(8), 2113–2117.

Ma, J.K., Hikmat, B.Y., Wycoff, K., Vine, N.D., Chargelegue, D., Yu, L., Hein, M.B. and Lehner, T. (1998) Characterization of a recombinant plant monoclonal secretory antibody and preventive immunotherapy in humans. *Nature Medicine* 4(5), 601–606.

Marillonnet, S., Giritch, A., Gils, M., Kandzia, R., Klimyuk, V. and Gleba, Y. (2004) In planta engineering of viral RNA replicons: efficient assembly by recombination of DNA modules delivered by Agrobacterium. *Proceedings of the National Academy of Sciences USA* 101(18), 6852–6857.

McCormick, A.A., Kumagai, M.H., Hanley, K., Turpen, T.H., Hakim, I., Grill, L.K., Tuse, D., Levy, S. and Levy, R. (1999) Rapid production of specific vaccines for lymphoma by expression of the tumor-derived single-chain Fv epitopes in tobacco plants. *Proceedings of the National Academy of Sciences USA* 96(2), 703–708.

McCormick, A.A., Reinl, S.J., Cameron, T.I., Vojdani, F., Fronefield, M., Levy, R. and Tuse, D. (2003) Individualized human scFv vaccines produced in plants: humoral anti-idiotype responses in vaccinated mice confirm relevance to the tumor Ig. *Journal of Immunology Methods* 278(1–2), 95–104.

McCormick, A.A., Reddy, S., Reinl, S.J., Cameron, T.I., Czerwinksci, D.K., Vojdani, F., Hanley, K.M., Garger, S.J., White, E.L., Novak, J., Barrett, J., Holtz, R.B., Tusac, D. and Levy, R. (2008) Plant-produced idiotype vaccines for the treatment of non-Hodgkin's lymphoma: Safety and immunogenicity in a phase I clinical study. *Proceedings of the National Academy of Sciences USA* 105(29), 10131–10136.

Mullard, A. (2012) Can next-generation antibodies offset biosimilar competition? *Nature Reviews - Drug Discovery* 11(6), 426–428.

Musiychuk, K., Stevenson, N., Bi, H., Farrance, C.E., Orozovic, C., Brodelius, M., Brodelius, P., Horsey, A., Ugulava, N., Shamloul, A.-M., Mett, V., Rabindran, S., Streatfield, S.J. and Yusibov, V. (2007) A launch vector for the production of vaccine antigens in plants. *Influenza* 1, 19–25.

Olinger, G.G., Pettitt, J., Kim, D., Working, C., Bohorov, O., Bratcher, B., Hiatt, E., Hume, S.D., Johnson, A.K., Morton, J., Pauly, M., Whaley, K.J., Lear, C.M., Biggins, J.E., Scully, C., Hensley, L. and Zeitlin, L. (2012) Delayed treatment of Ebola virus infection with plant-derived monoclonal antibodies provides protection in rhesus macaques. *Proceedings of the National Academy of Sciences USA* 109(44), 18030–18045.

Panella, N.A., Burkhalter, K.L., Langevin, S.A., Brault, A.C., Schooley, L.M., Biggerstaff, B.J., Nasci, R.S. and Komar, N. (2005) Rapid West Nile virus antigen detection. *Emerging Infectious Disease* 11(10), 1633–1635.

Peeters, K., De Wilde, C. and Depicker, A. (2001) Highly efficient targeting and accumulation of a Fab fragment within the secretory pathway and apoplast of *Arabidopsis thaliana*. *European Journal of Biochemistry* 268(15), 4251–4260.

Phoolcharoen, W., Bhoo, S.H., Lai, H., Ma, J., Arntzen, C.J., Chen, Q. and Mason, H.S. (2011a) Expression of an immunogenic Ebola immune complex in *Nicotiana benthamiana*. *Plant Biotechnology Journal* 9(7), 807–816.

Phoolcharoen, W., Dye, J.M., Kilbourne, J., Piensook, K., Pratt, W.D., Arntzen, C.J., Chen, Q., Mason, H.S. and Herbst-Kralovetz, M.M. (2011b) A nonreplicating subunit vaccine protects mice against lethal Ebola virus challenge. *Proceedings of the National Academy of Sciences USA* 108(51), 20695–20700.

Platis, D. and Labrou, N.E. (2008) Affinity chromatography for the purification of therapeutic proteins from transgenic maize using immobilized histamine. *Journal of Separation Science* 31(4), 636–645.

Pogue, G., Vojdani, F., Palmer, K.C., Hiatt, E., Hume, S., Phelps, J., Long, L., Bohorova, N., Kim, D., Pauly, M., Velasco, J., Whaley, K., Zeitlin, L., Garger, S., White, E., Bai, Y., Haydon, H. and Bratcher, B. (2010) Production of pharmaceutical-grade recombinant aprotinin and a monoclonal antibody product using plant-based transient expression systems. *Plant Biotechnology Journal* 8(5), 638–654.

Porta, C., Spall, V.E., Findlay, K.C., Gergerich, R.C., Farrance, C.E. and Lomonossoff, G.P. (2003) Cowpea mosaic virus-based chimaeras. Effects of inserted peptides on the phenotype, host range, and transmissibility of the modified viruses. *Virology* 310(1), 50–63.

Pushko, P., Bray, M., Ludwig, G.V., Parker, M., Schmaljohn, A., Sanchez, A., Jahrling, P.B. and Smith, J.F. (2000) Recombinant RNA replicons derived from attenuated Venezuelan equine encephalitis virus protect guinea pigs and mice from Ebola hemorrhagic fever virus. *Vaccine* 19(1), 142–153.

Regnault, A., Lankar, D., Lacabanne, V., Rodriguez, A., Thery, C., Rescigno, M., Saito, T., Verbeek, S., Bonnerot, C., Ricciardi-Castagnoli, P. and Amigorena, S. (1999) Fcgamma receptor-mediated induction of dendritic cell maturation and major histocompatibility complex class I-restricted antigen presentation after immune complex internalization. *Journal of Experimental Medicine* 189(2), 371–380.

Richardson, J. S., Yao, M.K., Tran, K.N., Croyle, M.A., Strong, J.E., Feldmann, H. and Kobinger, G.P. (2009) Enhanced protection against Ebola virus mediated by an improved adenovirus-based vaccine. *PLoS ONE* 4(4), e5308.

Rivera, A., Gomez-Lim, M., Fernandez, F. and Loske, A. (2012) Physical methods for genetic plant transformation. *Physics of Life Reviews* 9(3), 308–345.

Roque, A.C.A., Lowe, C.R. and Taipa, M.A. (2004) Antibodies and genetically engineered related molecules: production and purification. *Biotechnology Progress* 20(3), 639–654.

Sainsbury, F. and Lomonossoff, G.P. (2008) Extremely high-level and rapid transient protein production in plants without the use of viral replication. *Plant Physiology* 148(3), 1212–1218.

Sainsbury, F., Sack, M., Stadlmann, J., Quendler, H., Fischer, R. and Lomonossoff, G.P. (2010) Rapid transient production in plants by replicating and non-replicating vectors yields high quality functional anti-HIV antibody. *PLoS ONE* 5(11), e13976.

Samuel, M.A., Wang, H., Siddharthan, V., Morrey, J.D. and Diamond, M.S. (2007) Axonal transport mediates West Nile virus entry into the central nervous system and induces acute flaccid paralysis. *Proceedings of the National Academy of Sciences USA* 104(43), 17140–17145.

Sanford, J.C. (1990) Biolistic plant transformation. *Physiologia Plantarum* 79(1), 206–209.

Santi, L., Batchelor, L., Huang, Z., Hjelm, B., Kilbourne, J., Arntzen, C.J., Chen, Q. and Mason, H.S. (2008) An efficient plant viral expression system generating orally immunogenic Norwalk virus-like particles. *Vaccine* 26(15), 1846–1854.

Scholthof, H.B., Morris, T.J. and Jackson, A.O. (1993) The capsid protein gene of tomato bushy stunt virus is dispensable for systemic movement and can be replaced for localized expression of foreign genes. *Molecular Plant-Microbe Interactions* 6, 309–322.

Strasser, R., Stadlmann, J., Schahs, M., Stiegler, G., Quendler, H., Mach, L., Glössl, J., Weterings, K., Pabst, M. and Steinkellner, H. (2008) Generation of glyco-engineered *Nicotiana benthamiana* for the production of monoclonal antibodies with a homogeneous human-like N-glycan structure. *Plant Biotechnology Journal* 6(4), 392–402.

Strasser, R., Castilho, A., Stadlmann, J., Kunert, R., Quendler, H., Gattinger, P., Jez, J., Rademacher, T., Altmann, F., Mach, L. and Steinkellner, H. (2009) Improved virus neutralization by plant-produced anti-HIV antibodies with a homogeneous b1,4-galactosylated N-glycan profile. *Journal of Biological Chemistry* 284(31), 20479–20485.

Thuenemann, E.C., Meyers, A.E., Verwey, J., Rybicki, E.P. and Lomonossoff, G.P. (2013) A method for rapid production of heteromultimeric protein complexes in plants: assembly of protective bluetongue virus-like particles. *Plant Biotechnology Journal* 11(7), 839–846.

Umana, P., Jean-Mairet, J., Moudry, R., Amstutz, H. and Bailey, J.E. (1999) Engineered glycoforms of an antineuroblastoma IgG1 with optimized antibody-dependent cellular cytotoxic activity. *Nature Biotechnology* 17(2), 176–180.

Whaley, K., Morton, J., Hume, S., Hiatt, E., Bratcher, B., Klimyuk, V., Hiatt, A., Pauly, M. and Zeitlin, L. (2012) *Emerging Antibody-based Products*. Springer, Berlin, Heidelberg, pp. 1–20.

Wilson, J.A., Hevey, M., Bakken, R., Guest, S., Bray, M., Schmaljohn, A.L. and Hart, M.K. (2000) Epitopes involved in antibody-mediated protection from Ebola virus. *Science* 287(5458), 1664–1666.

Wilson, J.A., Bosio, C.M. and Hart, M.K. (2001) Ebola virus: the search for vaccines and treatments. *Cellular and Molecular Life Sciences* 58(12–13), 1826–1841.

Yin, J., Li, G., Ren, X. and Herrler, G. (2007) Select what you need: a comparative evaluation of the advantages and limitations of frequently used expression systems for foreign genes. *Journal of Biotechnology* 127(3), 335–347.

Zeitlin, L., Pettitt, J., Scully, C., Bohorova, N., Kim, D., Pauly, M., Hiatt, A., Ngo, L., Steinkellner, H., Whaley, K.J. and Olinger, G.G. (2011) Enhanced potency of a fucose-free monoclonal antibody being developed as an Ebola virus immunoprotectant. *Proceedings of the National Academy of Sciences USA* 108(51), 20690–20694.

Zeitlin, L., Bohorov, O., Bohorova, N., Hiatt, A., Kim, D.H., Pauly, M.H., Velasco, J., Whaley, K.J., Barnard, D.L., Bates, J.T., Crowe, J.E., Piedra, P.A. and Gilbert, B.E. (2013) Prophylactic and therapeutic testing of *Nicotiana*-derived RSV-neutralizing human monoclonal antibodies in the cotton rat model. *mAbs* 5(2), 263–269.

5 Plant-produced Virus-like Particles

Ann E. Meyers,[1]* Edward P. Rybicki[1,2] and Inga I. Hitzeroth[1]

[1]*Department of Molecular and Cell Biology, University of Cape Town, Rondebosch, South Africa;* [2]*Institute of Infectious Disease and Molecular Medicine, University of Cape Town, Cape Town, South Africa*

5.1 Introduction

The burden of disease in developing countries is disproportionately large compared to developed countries, mainly as a result of poor sanitation and lack of health services and infrastructure. There is therefore often a requirement by developing countries for not only large quantities of costly vaccines, but also for vaccines acting against diseases which are region-specific and which may not necessarily be of priority in developed countries, as well as vaccines that are readily available in the event of outbreaks or pandemics. Since most vaccines are manufactured in developing countries, their supply to undeveloped countries is expensive, slow and they are often difficult to obtain. There is a need for alternative methods of vaccine production for developing countries, as well as perhaps even an opportunity to explore alternative vaccine designs. These could provide a more cost effective, more readily available means of relieving the health burden on both humans and animals as well as be readily available for administration in the event of outbreaks and pandemics.

5.2 Current Vaccines

Many modern vaccines are still only available in the form of live-attenuated or inactivated virus (Plotkin, 2005), and these have several disadvantages. In the case of live-attenuated forms there is always the risk of recombin-ation with wild-type viruses (particularly multi-segmented viruses) to form novel virulent strains, and the complicated process of producing inactivated virus as a vaccine introduces the possibility of incomplete inactivation, which has serious implications for vaccinated individuals (Chackerian, 2007).

These disadvantages have led to the development of recombinant subunit vaccines, which are preferable for a number of different reasons. They usually

* Corresponding author, E-mail: Ann.Meyers@uct.ac.za

© CAB International 2014. *Plant-derived Pharmaceuticals: Principles and Applications for Developing Countries* (ed. K.L. Hefferon)

consist of a subset of viral protein/s, and are safer for human use as they contain no viral genetic material and are therefore not potentially infectious. In addition, the production process does not require handling of infectious material.

There are different types of subunit vaccines: those consisting of recombinant monomeric antigens, and multimeric particles more commonly known as virus-like particles (VLPs) or chimeric VLPs. These tend to be more stable than the soluble monomeric antigens and have been shown to elicit much stronger immune responses.

5.3 VLP Vaccines

Virus-like particles (VLPs) are formed by the self-assembly of a subset of viral-encoded proteins. They resemble the mature viral particles in size and shape but lack the viral genome. They have been shown to stimulate both the humoral and cellular arms of the immune system (Grgacic and Anderson, 2006). The repetitive nature of protein epitopes exposed on the surfaces of VLPs means they stimulate a strong antibody response as a result of B cells recognizing specific repetitive units (Bachmann *et al.*, 1993; Fehr *et al.*, 1998; Jegerlehner *et al.*, 2002). Moreover, their ability to be taken up by dendritic cells as a result of their similarity in size to wild-type virus particles and subsequent presentation by both MHC class I and class II molecules, renders them able to stimulate the cellular immune response as well (Chackerian, 2007).

There are a number of other advantages to using VLPs as vaccine candidates. Their multivalency reduces the concentration at which B cell activation takes place, compared to the amount of monomeric antigens required for a similar response (Dintzis *et al.*, 1976). In the case of some VLPs (e.g. human papillomavirus – HPV) it has been shown that co-administration of an adjuvant is not required for the induction of a strong antibody response (Zhang *et al.*, 2000), thus reducing the vaccine dose costs. VLPs can be manipulated and constructed so as to induce immune responses against homologous as well as heterologous viruses, in the form of chimeric particles (Plummer and Manchester, 2011). Selection of specific viral proteins for the composition of the VLPs (Roldao *et al.*, 2010) can circumvent the problem that occurs with immunodominance of certain antigen proteins. VLPs can be used to broaden the protective ability of the vaccine with the inclusion of multiple epitopes, which would lower vaccine dose amounts and concomitant vaccine costs. In the case of animal vaccines, VLP vaccines can be designed so as to exclude markers used for diagnosis of viral infection (Liu *et al.*, 2012), allowing for the distinction between infected and vaccinated animals (DIVA): this is a very important requirement in areas affected by disease outbreaks such as in the EU. These factors have led to rapid advances in the field of VLP vaccine development and production.

5.4 Conventional VLP Production Methods

The production of recombinant VLPs has been successful in various different expression systems including bacterial, yeast, insect and mammalian cells and, more recently, in plants (Roldao *et al.*, 2010).

Bacterial cells have been used for recombinant protein production by fermentation for many years (Chen, 2012), but they lack a post-translational modification system similar to mammalian cells, running the risk of producing proteins that may be reduced in their ability to stimulate an appropriate mammalian immune response. There is also a high risk of the production process becoming contaminated with bacterial products or viruses, which may go largely undetected in the final product. Yeast cells are a fairly common VLP production system. These are currently used for the production of the commercially available HPV VLP vaccine Gardasil® (Lu *et al.*, 2011). However, they also have a slightly different glycosylation protocol to that of mammalian cells (Yusibov and Rabindran, 2008). Insect cells have been used extensively to produce numerous types of VLPs (Kushnir *et al.*, 2012). Although post-translational modification in this system is more similar to mammalian systems, this method of VLP production is extremely expensive and, like bacterial cells, poses risks of contamination. Lastly, mammalian cells are also used extensively for recombinant protein production, mainly because their post-translational modification system is capable of coping with the complex modification required for more complex proteins (Wurm, 2004; Grillberger *et al.*, 2009). However, production costs of this process are also high and there are safety concerns about contamination of recombinant protein with other mammalian products.

Despite the above-mentioned disadvantages of these various expression systems, a variety of recombinant VLPs produced in bacterial, yeast, insect and mammalian cells are in production or clinical development (Kushnir *et al.*, 2012). VLP vaccines targeting human diseases caused by hepatitis B virus (HBV) and HPV are commercially available (Kushnir *et al.*, 2012), and there is also a porcine circovirus type 2 VLP vaccine (Porcilis PCV®) for animals, which is now commercially available (Mena and Kamen, 2011). The disadvantages and high costs of production in the above-mentioned expression systems have led to an ongoing search for alternative safer and cheaper systems. The reduction in cost will in turn increase the opportunities for developing countries to have access to vaccines currently not within their financial reach.

5.5 Production of VLPs in Plants

Plants are becoming increasingly more acceptable as a system for the production of recombinant proteins (Rybicki, 2009, 2010), including specifically VLPs (Chen and Lai, 2013; Scotti and Rybicki, 2013). The process is currently facilitated using one of two different methods: (i) production by transgenic plants engineered to encode specific antigenic proteins in plant cells; and (ii) transient production by infiltration of recombinant

Agrobacterium sp. hosting an expression vector encoding specific antigenic proteins into plant leaves under vacuum. There are now many examples of VLPs that have been produced in plants, a few of which have reached the clinical development stage of phase I trials (Kushnir *et al.*, 2012).

Plants are considered a viable and an appealing prospect for VLP vaccine production for developed countries for several reasons: (i) VLP vaccines can be designed to target a wide range of diseases to which those in undeveloped countries are prone; (ii) the production system is generally considered to be more cost effective; (iii) vaccines made in plant cells are considered to be safer; (iv) production in plants is easily scalable without huge infrastructure investments; (v) there is the possibility that they can be made thermostable; (vi) there is the possibility that plant-produced VLPs can be used for oral immunization; and (vii) plant-produced vaccines can be made rapidly in response to pandemics. These considerations are discussed in more detail below.

5.6 Disease Targets in Developing Countries

The disease burden in undeveloped countries affects many more people and animals than in developed countries and embraces a much broader range of diseases. Examples of commonly occurring human diseases include hepatitis B, human immunodeficiency virus (HIV/AIDS), human papillomavirus-related disease, human influenza, rotavirus-gastroenteritis (RVGE)-associated disease, malaria, rabies and respiratory syncytical virus (RSV) infections, amongst others.

Many of these human diseases have an enormous negative impact on the socio-economy of developing countries. HIV impairs the workforce sector of populations; human influenza can give rise to pandemics. HPV is the causative agent for cervical cancer, the second most prevalent cancer of women worldwide: cervical cancer results in approximately in 260,000 deaths annually, with 80% of them occurring in developing countries. Many of the diseases such as RVGE-associated disease, malaria and RSV affect primarily children – the World Health Organization (WHO) has recommended that all countries associated with high rotavirus-gastroenteritis-associated mortality rates should include rotavirus vaccine in their national immunization programme (WHO, 2013). Even though there are vaccines or drugs available to treat some of these diseases, their expense and scarcity render them inaccessible to many developing country populations.

The welfare of animals in developing countries is equally important, as many people rely on them for survival as well as economic gain. Animal diseases that impact heavily on their wellbeing and the economy of the country include foot and mouth disease (FMD), bluetongue disease, Rift Valley fever (RVF), porcine circovirus-associated disease, Newcastle disease, avian influenza and chicken anaemia.

Viruses such as those that cause FMD and Newcastle disease are very contagious amongst domestic animals and poultry, respectively, requiring strict quarantining of animals in the case of outbreaks, or even culling of entire stocks to arrest the spread. In many cases, such as in FMD infections, post-weaning wasting syndrome in swine caused by porcine circovirus-2 (PCV2), bluetongue disease (Coetzee *et al.*, 2012), Newcastle disease and chicken anaemia, which affects newly hatched chicks, enormous stock losses are experienced. Others such as Rift Valley fever virus (Pepin *et al.*, 2012) and avian influenza are zoonotic, and not only cause enormous numbers of fatalities and economic loss, but also increase the chances of human infection.

Finally, many developing countries are prone to outbreaks caused by neglected diseases that are geographically unique to their regions. These include diseases such as Lassa fever, Marburg virus disease, Chikungunya fever and Ebola. Although these are very infectious and some can be treated with antiviral therapy, they often go undiagnosed and cause rapid fatalities. There are currently no vaccines available for treatment of these diseases.

5.7 VLP Vaccines Against Target Diseases

There are currently a large number of conventionally made (yeast, bacterial, insect or mammalian cell-produced) VLP vaccine candidates available on the market or in clinical development that target the human diseases mentioned above which impact heavily on the health burden of developed countries. They can be divided into homologous VLP vaccines and chimeric VLP vaccines. Success with these has led to the development of some plant-produced candidate VLP vaccines as well.

5.7.1 Homologous VLP vaccines

The first recombinant HBV VLP vaccine was produced in the 1980s with production of the particle-forming hepatitis B surface antigen (HBsAg) in yeast cells (McAleer *et al.*, 1984). There are now several commercial HBsAg VLP vaccines available (Kushnir *et al.*, 2012), made in either yeast or Chinese hamster ovary (CHO) cells (Pniewski, 2013). Although the cost of these has decreased with increased demand and there have been improvements in manufacturing procedures, there is still a requirement for three intramuscular injections and cold-chain distribution, both of which pose an economic barrier for those living in developing countries.

A large amount of research has been carried out on alternative and cheaper strategies or recombinant HBV vaccine production based on plant-derived VLPs (Pniewski, 2013). The first reports of plant-produced HBV VLPs for injection and oral vaccination were in the 1990s with the small subunit (S)-HBsAg made in transgenic plants (Thanavala *et al.*, 1995; Kapusta *et al.*, 1999). Production of medium (M)-HBsAg and large (L)-HBsAg proteins was also tested as they are highly immunogenic, but yields in plants were relatively low (Pniewski, 2013). The transient expression of hepatitis B core antigen (HBcAg) was more successful in producing large amounts of VLPs, which showed immunogenicity in mice (Huang, Z. *et al.*, 2005).

Numerous structural HIV-1 vaccine candidate proteins have also been tested in trials and there has been some progress in HIV-1 VLP vaccine development (Buonaguro *et al.*, 2013). However, only one HIV-1 p17/p24 VLP vaccine has been tested in a phase 1 trial (IAVI, 2013). In conventional expression systems, HIV-1 Gag structural proteins can assemble spontaneously into VLPs. Scotti *et al.* (2009) have taken this further and shown that HIV-1 Gag-encoded VLPs can be formed in transplastomic plants at very high levels. However, these remain to be tested for their immunogenicity.

Currently, two VLP vaccines are commercially available to prevent cervical cancer caused by HPV: Gardasil® made by Merck & Co, consisting of L1 structural protein of HPV serotypes 6, 11, 16 and 18 and produced in a yeast expression system; and Cervarix® made by GlaxoSmithKline, which also consists of L1 structural protein representing HPV serotypes 16 and 18. Both are very efficacious, however their main limitation in the context of the developing world is their very high cost. As an answer to this, the production of papillomavirus L1 VLPs as candidate vaccines in plants has been extensively studied, and VLPs have been shown to be efficacious and protect rabbits from cottontail rabbit papillomavirus infection after a live virus challenge (Kohl *et al.*, 2006). Other HPV-related VLP production has been reviewed recently (Scotti and Rybicki, 2013).

H5N1 influenza VLPs produced in insect cells comprising the haemagglutinin (HA), neuraminidase (NA) and matrix 1 (M1) proteins, have been tested in humans in phase 1/2 trials and shown to stimulate cross-reactive neutralizing antibodies (Khurana *et al.*, 2011). López-Macías *et al.* (2011) have likewise proven the safety and immunogenicity of an H1N1 VLP vaccine candidate in humans. Medicago Inc., a clinical stage biopharmaceutical company, have taken things a step further and produced an H5N1 influenza VLP candidate vaccine in plants, comprising the HA protein only. Phase 1 trials of this vaccine show safety and promising immunogenicity (Landry *et al.*, 2010). More recently, they have announced the production of a similarly-produced influenza H7N9 VLP vaccine candidate (Medicago, 2013).

There is only one commercially available VLP vaccine licensed for veterinary use – Porcilis PCV®, which is produced in insect cells – which targets porcine circovirus 2 (PCV2) (Mena and Kamen, 2011).

Despite the paucity of commercialized VLP vaccines, there are several conventionally-made VLP candidate vaccines targeting diseases affecting economically important animals, which have reached the animal study stage of development (Crisci *et al.*, 2012). These include FMD, bluetongue, avian influenza, rotavirus infections, Newcastle disease, norovirus infections, Rift Valley fever and chicken anaemia, amongst others.

In the case of FMDV, it was reported that VLPs could be formed in insect cells by the processing of the FMDV P12A polyprotein into its three structural components (Cao *et al.*, 2009). These were shown to induce FMDV-specific antibodies and neutralizing antibodies in guinea pigs although levels were lower than those elicited by the commercial vaccine. Dus Santos *et al.* (2005) showed that VLPs composed of FMDV VP1 and produced in transgenic lucerne induced an immune response in mice when tested.

Extensive successful research has been done on the use of Bluetongue virus (BTV) VLPs as a protective vaccine for ruminants (Stewart *et al.*, 2012). Expression of the four BTV capsid proteins (VP2, VP3, VP5 and VP7) in *N. benthamiana* results in the formation of VLPs (Fig. 5.1). Thuenemann *et al.* (2013) have shown that such BTV VLPs expressed in plants can protect sheep from a live viral challenge.

In the case of Newcastle disease, a VLP-based candidate for poultry has been produced in avian cells and shown to be immunogenic in mice (McGinnes *et al.*, 2010). Although the first plant-made vaccine for which regulatory approval was granted was a recombinant haemagglutinin-neuraminidase protein against Newcastle disease, VLPs have not yet been produced in plants (Rybicki, 2010).

5.7.2 Chimeric VLP vaccines

A different means of expressing vaccine peptide epitopes or antigens transiently in plants is as surface fusions on particle-forming proteins or plant-virus particles. Plant virus particles such as tobacco mosaic virus

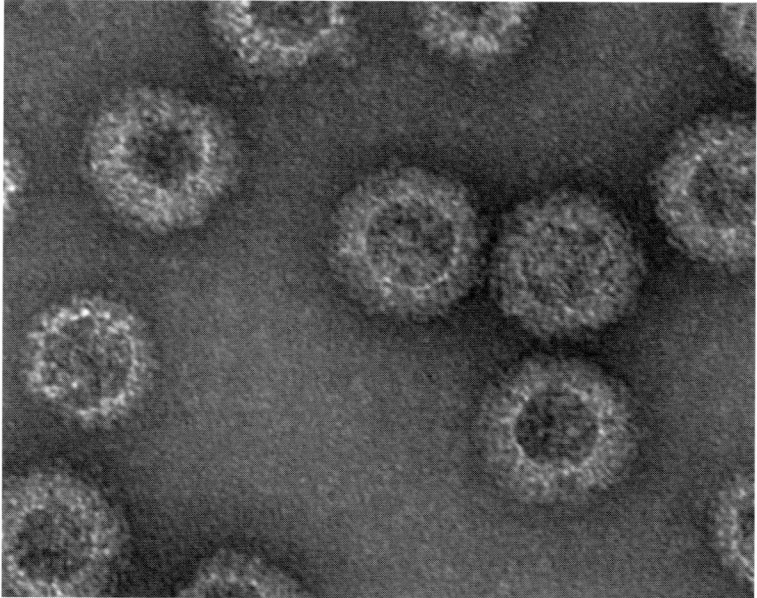

Fig. 5.1. Plant-produced BTV-8 VLPs.

(TMV), cowpea mosaic virus (CPMV), potato virus X (PVX) and bamboo mosaic virus (Scotti and Rybicki, 2013) have been used to express peptides from canine parvovirus, *Pseudomonas aeruginosa*, HIV, mink enteritis virus, FMDV and murine hepatitis virus, amongst others (Plummer and Manchester, 2011). Particle-forming proteins include HBsAg as well as the HBV core antigen (HBcAg), and HPV L1, which are used to express different viral peptides on their surfaces. Qian *et al.* (2008) expressed the HBV surface (S) protein fused to the HBV preS1 protein in rice seed. This generated VLPs that elicited a humoral response in mice, albeit fairly low. HBcAg has also been successfully used to express chimeric VLPs in transient systems. Huang, Y. *et al.* (2005) showed that a FMDV epitope VP1 could be fused to HBcAg and displayed on a surface loop, and was protective in mice. Another example is the fusion of the M2 protein of influenza virus fused to HBc-derived VLPs (Ravin *et al.*, 2012).

HPV-16 L1 VLPs have also been developed as vehicles for the display of different epitopes such as HIV-1 epitopes as well as the inclusion of HPV L2 to broaden the range of cross-type neutralizing antibodies against HPV (Varsani *et al.*, 2003; Schellenbacher *et al.*, 2009); chimeric HPV L1 VLPs have been made displaying a B cell epitope from HIV-1 gp41 envelope glycoprotein and shown to stimulate a humoral response in mice (Slupetzky *et al.*, 2001).

HIV-1 Pr55Gag VLPs are considered to be excellent immunogens as they stimulate both the humoral and cellular immune responses. HIV-1 Gag VLPs have also been used to generate chimeric particles (Halsey *et al.*, 2008; Scotti *et al.*, 2009, 2010).

Cucumber mosaic virus has been used to make chimeric hepatitis C virus VLPs by the display of two epitopes that were shown to elicit an immune response in mice which was specific (Piazzolla *et al.*, 2005).

5.8 Cost-effectiveness of Plant-produced VLP Vaccines

In developing countries most vaccines are comparatively expensive as a result of importation costs and requirements for refrigeration storage and transportation. This results in the inclusion of a limited number of vaccines in their Extended Programmes of Immunisation (EPI) and the exclusion of others which are equally necessary for immunization of the general population. It is therefore important that vaccines for developing countries are made affordable so that there is wide coverage of immunization. It is possible that cheaper plant-produced VLPs could increase the chances of inclusion of additional vaccines into their EPIs or at least reduce their cost to the public in undeveloped countries.

Although the initial costs of setting up a Good Manufacturing Practice (GMP) facility for plant-produced vaccines may be high, the subsequent maintenance costs of running the process are much lower than those required for maintaining current conventionally produced vaccine manufacturers (Gleba *et al.*, 2005). This has led to much calculation and the prognosis that plant-produced VLP vaccines may be cheaper. In the long term, this fact may even encourage the establishment of local production facilities, which would provide immediate access to vaccines and extend the range over which the vaccine is supplied, giving the opportunity of reaching people who do not normally have access to such health treatments.

5.9 Safety of VLP Vaccines

Although the conventional production of subunit VLP vaccines is safer in that it excludes the use of working with live virus and therefore the requirement for high safety levels during production, there is the risk of VLP vaccines made in bacterial, yeast and mammalian cells becoming contaminated with products that could go undetected in quality control, and cause subsequent harm to humans and animals immunized with them. In the case of VLP vaccine production in mammalian cells, although purification methods are fairly stringent, there may be contamination with animal products such as toxins or infectious agents (Tiwari *et al.*, 2009), which increases the cost as a result of extra purification steps required for removal. Additionally, these can be very difficult to remove. However, in the case of plant-produced VLP vaccines, the risk of contamination with animal products is excluded by the nature of their production vehicle.

5.10 Scalability and Availability of VLP Vaccines

Plant protein production is also a process that is easily scalable compared to that of other expression systems that require much greater amounts of capital investment and expertise. Again, this may be another factor to encourage the local establishment of a vaccine production facility. *Nicotiana* sp. in particular have a high biomass (up to 100 t/ha) (Tremblay *et al.*, 2010).

5.11 Thermostable Vaccines

One of the major drawbacks of conventionally produced vaccines is their requirement for transport and storage at specific temperatures in order to maintain their efficacy. In the developing world where vaccines are transported over long distances, there is often a question as to whether vaccines have indeed been maintained under such conditions and whether they will be effective. In addition, refrigeration and its associated costs are high for developing countries. A more desirable vaccine would be one that can withstand variation in temperatures and does not require refrigeration, i.e. is thermostable.

Diminsky *et al.* (1999) have shown that Chinese hamster ovary-produced HBsAg particles are stable and retained their immunogenicity when they were lyophilized in the presence of an osmoprotectant: this augurs well for similarly treated plant-produced HBsAg particles. This would allow for transport and storage at ambient temperatures prior to re-suspension and administration. Pniewski *et al.* (2011) have demonstrated that lettuce-produced HBsAg particles, lyophilized and formulated into tablets, remained stable and induced a mucosal and systemic humoral response in mice. In addition, although immunogenicity has not yet been proven, they showed that HBsAg VLPs composed of the medium (M) and large (L) surface antigens produced in both tobacco and lettuce can withstand lyophilization (Pniewski *et al.*, 2012).

Another way to obviate the need for the cold chain during transport to their destination and storage would be to generate vaccines in seeds. A large number of antigens have been expressed in transgenic seeds such as rice, safflower and maize (Lau and Sun, 2009; Hayden *et al.*, 2012) within storage protein bodies, with the idea that the environment within the seed is protective against degradation. This would allow for transport and storage without the need for a cold chain as well as possible oral delivery, which is discussed later.

5.12 Oral Vaccination

Oral vaccination would simplify the general vaccination procedure. A large amount of work has been carried out on oral dosing of plant-produced HBsAg particles (Shchelkunov and Shchelkunova, 2010). These were immunogenic in mice and humans dosed orally with various raw plant materials containing the recombinant VLPs; however, the effect was not long lasting in some cases and not very reproducible (Pniewski *et al.*, 2011).

For HBV, Pniewski (2013) suggests a novel vaccination regimen that combines the use of a plant-produced HBV antigen (HBsAg VLPs) which has been partially purified as an oral booster after a primary parenteral injection. Thanavala *et al.* (2005) have shown that oral administration of plant-produced HBsAg to mice as a booster elevated their immune response.

The problem with raw materials such as tobacco and potato is that they are unpalatable and difficult to digest. In addition, large doses of antigen are required for oral administration compared to parenteral administration as a result of degradation as the vaccine is transported through the gut system. Hayden *et al.* (2012) have addressed these issues by producing HBsAg antigen in maize seeds. These are digestible, provide an environment for minimization of protein degradation and apparently produce large quantities of recombinant protein (Stoger *et al.*, 2005). Hayden *et al.* (2012) showed that when the antigen was used as an oral boost to a primary injection of commercially produced HBV vaccine in mice, the mucosal response is amplified after the boost. Although VLP formation in the storage bodies of the maize seed was not shown, the formation of dimerized HBsAg was proven and suggests that particles were formed. However, this is certainly a possibility for the production of VLP vaccines not only for ease of administration but also reducing the cost as it does not require purification from the seed.

5.13 Ability to Respond to Pandemics

Perhaps one of the most important issues with vaccines is being able to provide sufficient doses in sufficient quantities, globally and swiftly, in response to pandemics. Developing countries tend to be left off the supply list as they are often remote and have to rely on imported vaccines as they do not manufacture their own. The influenza pandemic in 2009 indicated that the egg-based inactivated/attenuated viral vaccine production cannot cope under such pandemic conditions (López-Macías, 2012). Recently it has been shown that influenza VLPs produced in insect cells (Haynes, 2009) are as safe, and as efficient in stimulating immunity in animal models (Kang *et al.*, 2009) as conventional vaccines. More recently, it has been demonstrated that plant-produced influenza VLPs consisting of the HA protein only can be reliably made (D'Aoust *et al.*, 2008, 2010). Not only did these VLPs prevent infection in challenged ferrets but phase 1 clinical studies in humans indicated tolerance to the dose as well as acceptable seroconversion and seroprotection (Landry *et al.*, 2010). The most important aspect, however, is that the technology enabled the manufacture of this VLP vaccine to be developed within 3 weeks of the release of the influenza sequencing information (Landry *et al.*, 2010). In May 2013, Medicago Inc. announced that they successfully made in less than 1 month a VLP vaccine in plants for the potential pandemic influenza strain H7N9, which is responsible for the current outbreak of influenza in China (Medicago, 2013).

5.14 Conclusions

The processes involved in making VLP vaccines in plants have been improved to the extent that there are at least six examples that are at the clinical development stage. Collectively, the advantages of making plant-produced vaccines over conventional ones should make them more globally accessible. These would therefore have more application in developing countries, which could encourage a better coverage of vaccine use and perhaps even provide opportunities for their inclusion in extended immunization programmes (Rybicki *et al.*, 2013). In this context, plant-produced VLPs in particular can be used to address the 2015 Millennium Development Goals for improving several health aspects in developing countries, in the form of provision of vaccines for the treatment of both humans and animals (Penney *et al.*, 2011).

References

Bachmann, M.F., Rohrer, U.H., Kundig, R.M., Furki, K., Hengartner, H. and Zinkernagel, R.M. (1993) The influence of antigen organization on B cell responsiveness. *Science* 262, 1448–1451.
Buonaguro, L., Tagliamonte, M., Visciano, M.L., Tornesello, M.L. and Buonaguro, F.M. (2013) Developments in virus-like particle-based vaccines for HIV. *Expert Review of Vaccines* 12, 119–127.

Cao, Y., Lu, Z., Sun, J., Bai, X., Sun, P., Bao, H., Chen, Y., Guo, J., Li, D., Liu, X. and Liu, Z. (2009) Synthesis of empty capsid-like particles of Asia I foot-and-mouth disease virus in insect cells and their immunogenicity in guinea pigs. *Veterinary Microbiology* 137, 10–17.

Chackerian, B. (2007) Virus-like particles: flexible platforms for vaccine development. *Expert Review of Vaccines* 6, 381–390.

Chen, Q. and Lai, H. (2013) Plant-derived VLPs as vaccines. *Human Vaccines and Immunotherapeutics* 9, 26–49.

Chen, R. (2012) Bacterial expression systems for recombinant protein production: *E. coli* and beyond. *Biotechnology Advances* 30, 1102–1107.

Coetzee, P., Stokstad, M., Venter, E., Myrmel, M. and Van Vuuren, M. (2012) Bluetongue: a historical and epidemiological perspective with the emphasis on South Africa. *Virology Journal* 9, 198.

Crisci, E., Barcena, J. and Montoya, M. (2012) Virus-like particles: the new frontier of vaccines for animal viral infections. *Veterinary Immunology and Immunopathology* 148, 211–225.

D'Aoust, M.-A., Lavoie, P.-O., Couture, M.M.J., Trépanier, S., Guay, J.-M., Dargis, M., Mongrand, S., Landry, N., Ward, B.J. and Vézina, L.-P. (2008) Influenza virus-like particles produced by transient expression in *Nicotiana benthamiana* induce a protective immune response against a lethal viral challenge in mice. *Plant Biotechnology Journal* 6, 930–940.

D'Aoust, M.-A., Couture, M.M.J., Charland, N., Trépanier, S., Landry, N., Ors, F. and Vézina, L.-P. (2010) The production of hemagglutinin-based virus-like particles in plants: a rapid, efficient and safe response to pandemic influenza. *Plant Biotechnology Journal* 8, 607–619.

Diminsky, D., Moav, N., Gorecki, M. and Barenholz, Y. (1999) Physical, chemical and immunological stability of CHO-derived hepatitis B surface antigen (HBsAg) particles. *Vaccine* 18, 3–17.

Dintzis, H.M., Dintzis, R.Z. and Vogelstein, B. (1976) Molecular determinants of immunogenicity: the immunon model of immune response. *Proceedings of the National Academy of Sciences USA* 73, 3671–3675.

Dus Santos, M.J., Carrillo, C., Ardila, F., Ríos, R.D., Franzone, P., Piccone, M.E., Wigdorovitz, A. and Borca, M.V. (2005) Development of transgenic alfalfa plants containing the foot and mouth disease virus structural polyprotein gene P1 and its utilization as an experimental immunogen. *Vaccine* 23, 1838–1843.

Fehr, T., Skrastina, D., Pumpens, P. and Zinkernagel, R.M. (1998) T cell-independent type I antibody response against B cell epitopes expressed repetitively on recombinant virus particles. *Proceedings of the National Academy of Sciences USA* 95, 9477–9481.

Gleba, Y., Klimyuk, V. and Marillonnet, S. (2005) Magnifection – a new platform for expressing recombinant vaccines in plants. *Vaccine* 23, 2042–2048.

Grgacic, E.V.L. and Anderson, D.A. (2006) Virus-like particles: passport to immune recognition. *Methods* 40, 60–65.

Grillberger, L., Kreil, T.R., Nasr, S. and Reiter, M. (2009) Emerging trends in plasma-free manufacturing of recombinant protein therapeutics expressed in mammalian cells. *Biotechnology Journal* 4, 186–201.

Halsey, R.J., Tanzer, F.L., Meyers, A., Pillay, S., Lynch, A., Shephard, E., Williamson, A.L. and Rybicki, E.P. (2008) Chimaeric HIV-1 subtype C Gag molecules with large in-frame C-terminal polypeptide fusions form virus-like particles. *Virus Research* 133, 259–268.

Hayden, C.A., Streatfield, S.J., Lamphear, B.J., Fake, G.M., Keener, T.K., Walker, J.H., Clements, J.D., Turner, D.D., Tizard, I.R. and Howard, J.A. (2012) Bioencapsulation of the hepatitis B surface antigen and its use as an effective oral immunogen. *Vaccine* 30, 2937–2942.

Haynes, J.R. (2009) Influenza virus-like particle vaccines. *Expert Review of Vaccines* 8, 435–445.

Huang, Y., Liang, W., Wang, Y., Zhou, Z., Pan, A., Yang, X., Huang, C., Chen, J. and Zhang, D. (2005) Immunogenicity of the epitope of the foot-and-mouth disease virus fused with a hepatitis B core protein as expressed in transgenic tobacco. *Viral Immunology* 18, 668–677.

Huang, Z., Elkin, G., Maloney, B.J., Beuhner, N., Arntzen, C.J., Thanavala, Y. and Mason, H.S. (2005) Virus-like particle expression and assembly in plants: hepatitis B and Norwalk viruses. *Vaccine* 23, 1851–1858.

IAVI (2013) IAVIReport – clinical trials database. Available at: http://www.iavireport.org/trials-database/pages/default.aspx (accessed 15 April 2013).

Jegerlehner, A., Storni, T., Lipowsky, G., Schmid, M., Pumpens, P. and Bachmann, M.F. (2002) Regulation of IgG antibody responses by epitope density and CD21-mediated costimulation. *European Journal of Immunology* 32, 3305–3314.

Kang, S.M., Pushko, P., Bright, R.A., Smith, G. and Compans, R.W. (2009) Influenza virus-like particles as pandemic vaccines. In: Compans, R.W. and Orenstein, W.A. (eds) *Vaccines for Pandemic Influenza.* Springer, Berlin Heidelberg.

Kapusta, J., Modelska, A., Figlerowicz, M., Pniewski, T., Letellier, M., Lisowa, O., Yusibov, V., Koprowski, H., Pluciennczik, A. and Legocki, A.B. (1999) A plant-derived edible vaccine against hepatitis B virus. *The FASEB Journal* 13, 1796–1799.

Khurana, S., Wu, J., Verma, N., Verma, S., Raghunandan, R., Manischewitz, J., King, L.R., Kpamegan, E., Pincus, S., Smith, G., Glenn, G. and Golding, H. (2011) H5N1 virus-like particle vaccine elicits cross-reactive neutralizing antibodies that preferentially bind to the oligomeric form of influenza virus hemagglutinin in humans. *Journal of Virology* 85, 10945–10954.

Kohl, T., Hitzeroth, I., Stewart, D., Varsani, A., Govan, V.A., Christensen, N.D., Williamson, A.L. and Rybicki, E.P. (2006) Plant-produced cottontail rabbit papillomavirus L1 protein protects against tumor challenge: a proof-of-concept study. *Clinical and Vaccine Immunology* 13, 845–853.

Kushnir, N., Streatfield, S.J. and Yusibov, V. (2012) Virus-like particles as a highly efficient vaccine platform: diversity of targets and production systems and advances in clinical development. *Vaccine* 31, 58–83.

Landry, N., Ward, B.J., Trépanier, S., Montomoli, E., Dargis, M., Lapini, G. and Vézina, L.-P. (2010) Preclinical and clinical development of plant-made virus-like particle vaccine against avian H5N1 influenza. *PloS ONE* 5, e15559.

Lau, O.S. and Sun, S.S.M. (2009) Plant seeds as bioreactors for recombinant protein production. *Biotechnology Advances* 27, 1015–1022.

Liu, F., Ge, S., Li, L., Wu, X., Liu, Z. and Wang, Z. (2012) Virus-like particles: potential veterinary vaccine immunogens. *Research in Veterinary Science* 93, 553–559.

López-Macías, C. (2012) Virus-like particle (VLP)-based vaccines for pandemic influenza: Performance of a VLP vaccine during the 2009 influenza pandemic. *Human Vaccines and Immunotherapeutics* 8, 411–414.

López-Macías, C., Ferat-Osorio, E., Tenorio-Calvo, A., Isibasi, A., Talavera, J., Arteaga-Ruiz, O., Arriaga-Pizano, L., Hickman, S.P., Allende, M., Lenhard, K., Pincus, S., Connolly, K., Raghunandan, R., Smith, G. and Glenn, G. (2011) Safety and immunogenicity of a virus-like particle pandemic influenza A (H1N1) 2009 vaccine in a blinded, randomized, placebo-controlled trial of adults in Mexico. *Vaccine* 29, 7826–7834.

Lu, B., Kumar, A., Castellsague, X. and Giuliano, A. (2011) Efficacy and Safety of Prophylactic Vaccines against Cervical HPV Infection and Diseases among Women: A Systematic Review and Meta-Analysis. *BMC Infectious Diseases* 11, 13.

McAleer, W.J., Buynak, E.B., Maigetter, R.Z., Wampler, D.E., Miller, W.J. and Hilleman, M.R. (1984) Human hepatitis B vaccine from recombinant yeast. *Nature* 307, 178–180.

McGinnes, L.W., Pantua, H., Laliberte, J.P., Gravel, K.A., Jain, S. and Morrison, T.G. (2010) Assembly and biological and immunological properties of Newcastle disease virus-like particles. *Journal of Virology* 84, 4513–4523.

Mena, J.A. and Kamen, A.A. (2011) Insect cell technology is a versatile and robust vaccine manufacturing platform. *Expert Review of Vaccines* 10, 1063–1081.

News Medical (2013) New VLP vaccine candidate produced for H7N9 virus. Available at: http://www.news-medical.net/news/20130508/New-VLP-vaccine-candidate-produced-for-H7N9-virus.aspx (accessed 22 September 2014).

Penney, C.A., Thomas, D.R., Deen, S.S. and Walmsley, A.M. (2011) Plant-made vaccines in support of the Millennium Development Goals. *Plant Cell Reports* 30, 789–798.

Pepin, M., Laaberki, M.H., Dupinay, T., Marianneau, P. and Legras-Lachuer, C. (2012) Increasing importance of Bunyaviridae in public and veterinary health illustrated by hantaviruses, and the Schmallenberg and Rift Valley fever viruses. *Bulletin De L'Academie Veterinaire De France* 165, 339–346.

Piazzolla, G., Nuzzaci, M., Tortorella, C., Panella, E., Natilla, A., Boscia, D., De Stradis, A., Piazzolla, P. and Antonaci, S. (2005) Immunogenic properties of a chimeric plant virus expressing a hepatitis C virus (HCV)-derived epitope: new prospects for an HCV vaccine. *Journal of Clinical Immunology* 25, 142–152.

Plotkin, S.A. (2005) Vaccines: past, present and future. *Nature Medicine* 11(4), S5–S11.

Plummer, E.M. and Manchester, M. (2011) Viral nanoparticles and virus-like particles: platforms for contemporary vaccine design. *Wiley Interdisciplinary Reviews: Nanomedicine and Nanobiotechnology* 3, 174–196.

Pniewski, T. (2013) The twenty-year story of a plant-based vaccine against hepatitis B: stagnation or promising prospects? *International Journal of Molecular Sciences* 14, 1978–1998.

Pniewski, T., Kapusta, J., Bociąg, P., Wojciechowicz, J., Kostrzak, A., Gdula, M., Fedorowicz-Strońska, O., Wójcik, P., Otta, H., Samardakiewicz, S., Wolko, B. and Płucienniczak, A. (2011) Low-dose oral immunization with lyophilized tissue of herbicide-resistant lettuce expressing hepatitis B surface antigen for prototype plant-derived vaccine tablet formulation. *Journal of Applied Genetics* 52, 125–136.

Pniewski, T., Kapusta, J., Bociąg, P., Kostrzak, A., Fedorowicz-Strońska, O., Czyż, M., Gdula, M., Krajewski, P., Wolko, B. and Płucienniczak, A. (2012) Plant expression, lyophilisation and storage of HBV medium and large surface antigens for a prototype oral vaccine formulation. *Plant Cell Reports* 31, 585–595.

Qian, B., Shen, H., Liang, W., Guo, X., Zhang, C., Wang, Y., Li, G., Wu, A., Cao, K. and Zhang, D. (2008) Immunogenicity of recombinant hepatitis B virus surface antigen fused with preS1 epitopes expressed in rice seeds. *Transgenic Research* 17, 621–631.

Ravin, N.V., Kotlyarov, R.Y., Mardanova, E.S., Kuprianov, V.V., Migunov, A.I., Stepanova, L.A., Tsybalova, L.M., Kiselev, O.I. and Skryabin, K.G. (2012) Plant-produced recombinant influenza vaccine based on virus-like HBc particles carrying an extracellular domain of M2 protein. *Biochemistry (Moscow)* 77, 33–40.

Roldao, A., Mellado, M.C., Castilho, L.R., Carrondo, M.J. and Alves, P.M. (2010) Virus-like particles in vaccine development. *Expert Review of Vaccines* 9, 1149–1176.

Rybicki, E.P. (2009) Plant-produced vaccines: promise and reality. *Drug Discovery Today* 14, 16–24.

Rybicki, E.P. (2010) Plant-made vaccines for humans and animals. *Plant Biotechnology Journal* 8, 620–637.

Rybicki, E.P., Hitzeroth, I.I., Meyers, A., Dus Santos, M.J. and Wigdorovitz, A. (2013) Developing country applications of molecular farming: case studies in South Africa and Argentina. *Current Pharmaceutical Design* 19(31), 5612–5621.

Schellenbacher, C., Roden, R. and Kirnbauer, R. (2009) Chimeric L1-L2 virus-like particles as potential broad-spectrum human papillomavirus vaccines. *Journal of Virology* 83, 10085–10095.

Scotti, N. and Rybicki, E.P. (2013) Virus-like particles produced in plants as potential vaccines. *Expert Review of Vaccines* 12, 211–224.

Scotti, N., Alagna, F., Ferraiolo, E., Formisano, G., Sannino, L., Buonaguro, L., De Stradis, A., Vitale, A., Monti, L., Grillo, S., Buonaguro, F.M. and Cardi, T. (2009) High-level expression of the HIV-1 Pr55(gag) polyprotein in transgenic tobacco chloroplasts. *Planta* 229, 1109–1122.

Scotti, N., Buonaguro, L., Tornesello, M.L., Cardi, T. and Buonaguro, F.M. (2010) Plant-based anti-HIV-1 strategies: vaccine molecules and antiviral approaches. *Expert Review of Vaccines* 9, 925–936.

Shchelkunov, S.N. and Shchelkunova, G.A. (2010) Plant-based vaccines against human hepatitis B virus. *Expert Review of Vaccines* 9, 947–955.

Slupetzky, K., Shafti-Keramat, S., Lenz, P., Brandt, S., Grassauer, A., Sara, M. and Kirnbauer, R. (2001) Chimeric papillomavirus-like particles expressing a foreign epitope on capsid surface loops. *Journal of General Virology* 82, 2799–2804.

Stewart, M., Dovas, C.I., Chatzinasiou, E., Athmaram, T.N., Papanastassopoulou, M., Papadoulos, O. and Roy, P. (2012) Protective efficacy of Bluetongue virus-like and subvirus-like particles in sheep: presence of the serotype-specific VP2, independent of its geographic lineage, is essential for protection. *Vaccine* 30, 2131–2139.

Stoger, E., Ma, J.K.C., Fischer, R. and Christou, P. (2005) Sowing the seeds of success: pharmaceutical proteins from plants. *Current Opinion in Biotechnology* 16, 167–173.

Thanavala, Y., Yang, Y.F., Lyons, P., Mason, H.S. and Arntzen, C. (1995) Immunogenicity of transgenic plant-derived hepatitis-B surface-antigen. *Proceedings of the National Academy of Sciences USA* 92, 3358–3361.

Thanavala, Y., Mahoney, M., Pal, S., Scott, A., Richter, L., Natarajan, N., Goodwin, P., Arntzen, C.J. and Mason, H.S. (2005) Immunogenicity in humans of an edible vaccine for hepatitis B. *Proceedings of the National Academy of Sciences USA* 102, 3378–3382.

Thuenemann, E.C., Meyers, A.E., Verwey, J., Rybicki, E.P. and Lomonossoff, G.P. (2013) A method for rapid production of heteromultimeric protein complexes in plants: assembly of protective bluetongue virus-like particles. *Plant Biotechnology Journal* 11(7), 839–846.

Tiwari, S., Verma, P.C., Singh, P.K. and Tuli, R. (2009) Plants as bioreactors for the production of vaccine antigens. *Biotechnology Advances* 27, 449–467.

Tremblay, R., Wang, D., Jevnikar, A.M. and Ma, S. (2010) Tobacco, a highly efficient green bioreactor for production of therapeutic proteins. *Biotechnology Advances* 28, 214–221.

Varsani, A., Williamson, A.-L., De Villiers, D., Becker, I., Christensen, N.D. and Rybicki, E.P. (2003) Chimeric human papillomavirus type 16 (HPV-16) L1 particles presenting the common neutralizing epitope for the L2 minor capsid protein of HPV-6 and HPV-16. *Journal of Virolgy* 77, 8386–8393.

WHO (2013) Rotavirus vaccines. *Weekly Epidemiological Record*. WHO, Switzerland.

Wurm, F.M. (2004) Production of recombinant protein therapeutics in cultivated mammalian cells. *Nature Biotechnology* 22, 1393–1398.

Yusibov, V. and Rabindran, S. (2008) Recent progress in the development of plant derived vaccines. *Expert Review of Vaccines* 7, 1173–1183.

Zhang, L.F., Zhou, J., Chen, S., Cai, L.L., Bao, Q.Y., Zheng, F.Y., Lu, J.Q., Padmanabha, J., Hengst, K., Malcolm, K. and Frazer, I.H. (2000) HPV6b virus like particles are potent immunogens without adjuvant in man. *Vaccine* 18, 1051–1058.

6 Expression of the Capsid Protein of Human Papillomavirus in Plants as an Alternative for the Production of Vaccines

José Francisco Castillo Esparza,[1] Alberto Monroy García[2] and Miguel Angel Gómez Lim[1]*

[1]*Centro de Investigación y de Estudios Avanzados (CINVESTAV), Unidad Irapuato, Guanajuato, México;* [2]*Laboratorio de Inmunobiología, Unidad de Investigación en Diferenciación Celular y Cáncer, FES-Zaragoza, UNAM, México DF*

6.1 Introduction

Cervical cancer is the second most common cancer among women worldwide and it was the first cancer recognized by the WHO to be attributable to an infectious agent (Bosch *et al.*, 2002). Genital infection with human papillomavirus (HPV) is a common sexually transmitted disease and it poses a significant public health burden throughout the world, mainly in developing countries. HPV has also been established as a cause of cancer of the penis, anus, vagina and oropharyngeal tract (Monk and Tewari, 2007). The most efficient and cost-effective method to control infectious diseases has been the use of vaccination (Markowitz *et al.*, 2013). Therefore, an affordable HPV vaccine to prevent cancer has been a research priority for many years.

Human papillomaviruses are small, non-enveloped, double-stranded DNA viruses that infect human squamous and cutaneous epithelial cells. The high-risk types HPV-16 and HPV-18 are the most prevalent genotypes in invasive cervical cancers, accounting for approximately 70% of diagnosed cervical cancers worldwide and 300,000 deaths every year (Bruni *et al.*, 2010). It is predicted that if the risks and population growth remain constant, by 2020 there will be a 42% increase in cervical cancer cases if no novel preventive interventions are undertaken. The high incidence and cost to screen, treat and provide psychological support contribute to the large economic burden of the disease (Chih *et al.*, 2013).

The demonstration that the major component of the viral capsid protein L1 easily self-assembles to form virus-like particles (or VLPs) has allowed

* Corresponding author, E-mail: mgomez@ira.cinvestav.mx

the development of commercial vaccines. Two multivalent HPV L1 VLP-based prophylactic vaccines have been licensed and are highly effective in the prevention of vaccine type infections and associated disease. Gardasil® (Merck & Co., Inc.) contains L1 VLPs of low risk genital wart types 6 and 11 and high cancer risk types 16 and 18, produced in *Saccharomyces cerevisiae*. Cervarix® (GlaxoSmithKline Biologicals) contains L1 VLPs from types 16 and 18, produced via recombinant baculovirus in insect cells. However, their type restriction and high cost limit their widespread application, particularly in developing countries. Therefore, there is an urgent need for affordable second generation HPV vaccines that broaden protection to include multiple oncogenic HPV types. Furthermore, the commercial vaccines are useful in preventing infection by HPV but they are not useful for therapeutic purposes. This makes it necessary to develop therapeutic vaccines effective to counteract the disease.

In this chapter, we will analyse the different strategies that have been used for the expression and production of heterologous proteins in plants and the impact they have had in the field of production of vaccines against HPV.

6.2 Plants as Bioreactors for the Production of Vaccines

The use of plants for the expression of vaccines is an affordable and attractive alternative, especially when high quantities are required. Considerable progress has been made in the field since Charles Arntzen first envisaged the idea in the early 1990s. There are now a number of examples demonstrating the successful expression of subunit candidate vaccines both for humans and animals in plants (see Gómez Lim, 2011 for a review). These include antigens from bacterial and viral sources infecting humans, domestic or wild animals and representing secreted toxins and cell or viral surface antigens. Production of vaccines in plants has many advantages (Table 6.1): low cost, scalability, low health risks and the potential ability to be administered as unprocessed or partially processed material (Waheed *et al.*, 2012). Up-scaling of this production system can be achieved more easily compared to other systems such as mammalian cell culture, where up-scaling of the fermentation process leads to increasing production costs. In theory, the costs of an IgA expressed in plants are only 1–10% compared to the expression in hybridoma cells (Frenzel *et al.*, 2013).

Table 6.1. Advantages of plants as production platforms for recombinant proteins.

Low production cost
Rapid scaling up or down
Post-translational modifications
Flexible system allowing combinations of antigens
The plant material with the antigen is stable for years and requires no cold network for transport
No risk of contamination by human or animal pathogens

Over the last few years it has been possible to overcome some of the limitations of plants as bioreactors such as low yields and variable quality of products, together with an improved stability of proteins and more efficient manufacturing processes. There are currently four proteins derived from plants used for diagnosis commercially available. In May 2012, the US Food and Drug Administration (FDA) approved for use in humans the first compound produced in plants, an enzyme produced in carrot cells for the treatment of Gaucher disease and produced by the company Protalix Biotherapeutics. This represents a milestone in technology and will certainly help to get more products approved and marketed.

The first approach required the use of transgenic plants, but this technology has now been replaced by transient expression methods, which yield much higher levels of transgene expression, are more easily scalable and take considerably less time. The development of viral-based methods (Gómez Lim, 2011), which allow achieving very high yields of recombinant proteins in several days, has certainly paved the way for a number of applications.

On the other hand, the fact that various VLPs have been successfully produced as candidate vaccines in plants, opens the door for the production of low-cost, highly efficient plant-based HPV vaccines (Scotti and Rybicki, 2013).

6.3 Antigen Expression of the Human Papillomavirus in Plants

The history of HPV prophylactic vaccines began when it was shown that the major component of the viral capsid protein L1 easily self-assembled to form VLPs (Kirnbauer *et al.*, 1992). VLPs are more effective and safer than subunit vaccines, because they are conformationally more authentic than live attenuated vaccines and because they are free of viral genetic material (see Chapter 4, this volume). That is why this strategy has been extensively tested with HPV, which has no envelope and whose particles have been produced in various systems. The outer shell of the HPV particle is composed of 72 L1 pentamers, arranged on a T = 7 icosahedral lattice (Bishop *et al.*, 2007). Recombinant L1 has been shown to form T = 7 VLPs *in vivo* and *in vitro*. L1 can assemble *in vitro* into VLPs spontaneously using high salt or low pH conditions (Bishop *et al.*, 2007). L2, the minor capsid protein, is present at a proportion of one molecule for each L1 capsomere and is not required for VLP assembly. The viral genomic DNA is packaged within the L1/L2 capsid as a minichromosome (Zhou *et al.*, 1993).

Plants have been suggested as the ideal system for production of human and animal papilloma L1 HPV either as VLPs or capsomeres employing transient, transgenic or plastid expression (Waheed *et al.*, 2012; Scotti and Rybicki, 2013). Plants such as tobacco, potato, tomato and *Arabidopsis* have been employed but transient expression systems have been particularly useful for rapid production and significantly higher antigen levels have been obtained in comparison with stable nuclear transformation (Table 6.2). Depending on the antigen employed, plant-based vaccines against HPV have been for prophylactic and therapeutic uses.

6.4 Development of First-generation Plant-based Vaccines Against HPV

The aim of the original strategies for HPV prophylactic vaccination was the induction of neutralizing antibodies against L1 protein. The first published evidence that HPV L1 could be expressed in transgenic plants was a patent (Sohn *et al.*, 2002). The authors expressed L1 and L2 from HPV-16 and -18 in tobacco and tomato and claimed assembly of VLPs (although no evidence was provided) and immunogenicity of the preparations in mice immunized orally and intraperitoneally. However, this was never published in any other form.

The HPV L1 protein was expressed in plants for the first time in 2003 by three groups independently. The first group described expression of a human-codon optimized HPV-16 L1 cDNA in transgenic tobacco and potato and obtained a yield of approximately 12 mg kg^{-1} in potato tubers and 20 mg kg^{-1} in tobacco leaves (Biemelt *et al.*, 2003). The L1 protein assembled into VLPs in the plant cytoplasm but they showed no immunogenicity by oral route in mice. Nevertheless, the tobacco version was highly immunogenic when injected in a purified form. A second group expressed a plant-codon optimized HPV-11 L1 protein gene in transgenic potato tubers and obtained a yield of 20 µg kg^{-1}, which assembled into recognizable VLPs. Similar to the previous work, the VLPs were weakly immunogenic in mice by oral administration (Varsani *et al.*, 2003b). A third group also expressed a human-codon optimized HPV-16 L1 gene in transgenic tobacco and potato plants (Warzecha *et al.*, 2003). Expression levels ranged from 4×10^{-6} g kg^{-1} for HPV-16 L1 in the first case, through 20×10^{-6} g kg^{-1} for HPV-11 VLPs in potato, to 12×10^{-3} g kg^{-1} in potato in the second and 20×10^{-3} g kg^{-1} in tobacco for the optimized HPV-16 L1 gene in the third. While all of the products were immunogenic in experimental animals, oral immunization with HPV-11 and -16 products was only weakly immunogenic and yields were too low for meaningful production.

Plant-based expression was employed to demonstrate that rabbit HPV L1 expressing L2 epitopes on their surface could confer partial or complete protection against a viral challenge in rabbits (Palmer *et al.*, 2006). In further experiments, intramuscular injection of plant crude extracts containing the native L1 protein in the presence of Freund's adjuvant provided complete protection in rabbits against a challenge with rabbit HPV (Kohl *et al.*, 2006). In this last case, capsomeres but no VLPs were obtained and even though no neutralizing antibodies were detected, the animals were protected.

These same authors expressed the native HPV-11 L1 gene in transgenic tobacco and *Arabidopsis* plants (obtaining 2 and 12 mg kg^{-1} of L1 antigen,

Table 6.2. HPV antigens for vaccines expressed in plants.

Antigen and papillomavirus	Type of structure	Expression levels and system	Immunological data	Reference
L1 (VPH-16)	VLPs	12 mg kg^{-1} potato 20 mg kg^{-1} tobacco	Orally low immunogenicity. The version of tobacco was highly immunogenic when injected	Biemelt *et al.*, 2003
L1 (VPH-11)	VLPs	20 µg kg^{-1} potato	Orally low immunogenicity	Warzecha *et al.*, 2003
L1 (VPH-16)	VLPs	4 µg kg^{-1} tobacco	Low immunogenicity parenteral	Varsani *et al.*, 2003a
L1 (VPH-16)	VLPs	0.05% TSP tobacco	ND	Liu *et al.*, 2005
L1 (VPH-16)	Tobacco mosaic virus as vector	40 g kg^{-1} *Nicotiana benthamiana*	ND	Varsani *et al.*, 2006
Rabbit papillomavirus with L2 epitopes	VLPs	ND	Partial or complete protection against a challenge with the virus	Palmer *et al.*, 2006
L1 HPV rabbit	Capsomeres	0.4 mg kg^{-1} *N. benthamiana*	Full protection in the absence of detectable neutralizing antibodies	Kohl *et al.*, 2006
L1 (VPH-11)	VLPs Capsomeres Tobacco mosaic virus as vector	2 mg kg^{-1} tobacco 12 mg kg^{-1} *Arabidopsis* 10 mg kg^{-1} *N. benthamiana*	Seroconversion. Neutralizing antibodies were not detectable	Kohl *et al.*, 2007
L1 (VPB-16)	VLPs Capsomeres	380 mg kg^{-1} *N. benthamiana* tobacco	Neutralizing antibody seroconversion	Maclean *et al.*, 2007
L1 (HPV-16) plus the first 14 amino acids of the Rubisco or β-subunit of ATPase	VLPs	60 mg kg^{-1} in tobacco	ND	Lenzi *et al.*, 2008
L1 (VPH-16)	VLPs	3 g kg^{-1} in tobacco chloroplasts	Neutralizing antibody seroconversion	Jagu *et al.*, 2009
L2 (VPH-16)	PVX in *N. benthamiana*	170 mg kg^{-1} of fresh leaf tissue in *N. benthamiana*	Seroconversion	Cerovská *et al.*, 2012
L1(VPH-16)	Chimeric VLPs containing a string of T-cell epitopes from HPV-16 E6 and E7 fused to its C-terminus	Tomato 0.05–0.1% of TSP	cVLPs induced persistent IgG antibodies for over 12 months, with reactivity and neutralizing activity for VLPs composed of only the HPV-16 L1 protein	Monroy *et al.*, 2013
L1/L2 (HPV-16)	Chimeric VLPs	*N. benthamiana* ~1.2 g kg^{-1} plant tissue	Elicited anti-L1 and anti-L2 responses in mice, and anti-sera neutralized homologous HPV-16 and heterologous HPV-52 pseudovirions	Pineo *et al.*, 2013

respectively) and in recombinant tobacco mosaic virus in *Nicotiana tabacum* plants (with a yield of 10 mg kg^{-1}) (Kohl *et al.*, 2007). The authors were able to increase the expression levels in the transgenic plants by deleting the last 34 amino acids of the 3′ end, which code for the nuclear localization signal. Furthermore, the authors reported an increased degradation of L1 in tobacco but not in *Arabidopsis* or *N. tabacum*. As can be seen, the yield in all of these reports was below 20 mg kg^{-1} fresh weight (FW), which roughly corresponds to 0.5% of total soluble protein (TSP).

Maclean *et al.* (2007) performed a detailed study of the expression of HPV-16 L1 protein in *N. benthamiana*. By expressing different versions of the L1 gene (human, plant and native-codon genes) they found that the human version showed the higher yields in transient expression (380 mg kg^{-1}, 15% TSP) similar to chloroplast expression (>5 g kg^{-1}, 17% TSP). Parenteral immunization of mice with plant extracts were highly immunogenic but the addition of Freund's adjuvant did not increase the antibody titres. Interestingly, from T2 onwards, the transgene showed silencing in all transgenic plants. This has been a recurrent problem with transgenic L1 plants besides the low levels of expression.

Lenzi *et al.* (2008), employing a chloroplast-codon optimized or a native L1, were able to reach expression of 1.5% TSP, obtaining capsomeres as well as VLPs. They fused the 5′ end of the L1 gene with the first 14 amino acids of the ATPase β subunit or the large Rubisco subunit and deleted the nuclear localization signal. Fernández-San Millán *et al.* (2008) expressed the native HPV-16 L1 gene in tobacco chloroplasts and they obtained the highest yield yet reported so far for L1, 3 g kg^{-1} FW or 24% TSP. The protein self-assembled into VLPs, which were able to induce high titres of neutralizing antibodies in mice, intraperitoneally injected in the presence of Freund's adjuvant or aluminium hydroxide. Waheed *et al.* (2011a) successfully expressed a HPV-16 L1 protein modified to assemble only to capsomeres and obtained a yield of 1.5% TSP corresponding to about 0.18 g kg^{-1} FW. This same group fused a similar construct to the *Escherichia coli* heat-labile enterotoxin subunit B (LTB), as this potent adjuvant can increase immunogenicity. The levels of expression were slightly better than before (2% TSP or about 0.25 g kg^{-1} FW) and proper folding and display of conformational epitopes for both LTB and L1 in the fusion protein was confirmed by GM1-ganglioside binding assay and ELISA. Unfortunately, the plants presented a series of abnormal phenotypic effects (Waheed *et al.*, 2011b). LTB was also employed by Hongli *et al.* (2013), who expressed HPV16 L1 and LTB in transgenic tobacco plants. High levels of specific IgG and IgA (i.e. systemic and mucosal immune response) were induced when mice were immunized orally with L1 combined with LTB. Regnard *et al.* (2010), using a replicating bean yellow dwarf geminivirus-derived vector, expressed HPV-16 L1 in the cytoplasm obtaining yields of 0.55 g kg^{-1}. Since papillomaviruses also infect horses and cattle, an L1 protein from bovine papillomaviruses was expressed in *N. benthamiana* using the pEAQ-HT vector and high yields of VLP-assembled L1 protein (0.18 g kg^{-1}) were achieved (Love *et al.*, 2012).

Not only the L1 protein from HPV-16 has been expressed in plants. L1 from HPV-8 has also been successfully expressed in plants employing a native gene sequence (Matić *et al.*, 2012). HPV-8 is a high-risk cutaneous HPV suspected of human tumorigenesis and has also been found in non-melanoma skin cancer lesions, occurring in HIV-infected people. Reasonably high levels of expression were achieved with the TMV-derived Icon Genetics (Halle, Germany) vectors and with the pEAQ-HT vector in *N. benthamiana* (about 0.06 g kg^{-1}) and expression was enhanced fourfold by deletion of the nuclear localization signal.

6.5 Development of Second-generation Plant-based Vaccines Against HPV

The aim of second-generation vaccines is basically to reduce burden of disease or to alleviate the potential problem of disease succession. A general approach for cancer immunotherapy is the stimulation of the immune system against target antigens present in cancer cells. In cervical cancer, the immunodominant antigens are the proteins E6 and E7, which are necessary to initiate and maintain the tumorous state and for that reason they are expressed constitutively (Zur Hausen, 2000). Interaction of E6 and E7 with p53 and pRB (two important human tumour suppressor genes) results in loss of control of cell cycle, of apoptosis and of viral defence mechanisms. Therefore the possibility of using viral antigens as therapeutic vaccines has been very attractive as these proteins are from viral origin.

Several vaccines have been developed based on the use of E7 and E6 in the past decade including therapies with peptides/proteins, viral vectors, DNA vaccines, dendritic cells and chimeric VLPs. These vaccines have had a moderate success rate on induction of cellular immune responses to E6 and E7 in HPV-infected patients, however it has been demonstrated that immune responses do not always correlate with a clinical response (Münger *et al.*, 2004). An ideal vaccine against cervical cancer has to address the fact that viral infection creates an immunosuppressor environment. An efficient therapeutic vaccine against cervical cancer should induce effector T cells, acute inflammation at the site of tumour and overcome immunosuppression (Münger *et al.*, 2004). These requirements may be fulfilled by increasing the 'visibility' of the antigen.

There are few studies with viral oncoproteins expressed in plants (Table 6.3). The approach that many groups used at first was quite simple, the expression of a single HPV antigen in plant tissue and one of the primary targets was the E7 oncoprotein of HPV-16, which was the first HPV antigen expressed in plants. E7 was transiently expressed in leaves of *N. benthamiana*, *N. rustica*, *N. tabacum*, *Chenopodium quinoa* and *Lycopersicon esculentum* cultivar Micro-Tom, using potato virus X (PVX) as vector. The highest expression was obtained in *N. benthamiana* (3–4 µg g^{-1} FW). Crude extracts were used to immunize C57BL/6 mice, which developed a strong cellular and humoral immune response and a 40% protection was obtained after a

Table 6.3. HPV antigens expressed in plants for therapeutic vaccines.

HPV antigen	Production and performance system	Immunological data	Reference
VPH-16 E7	PVX vector in *Nicotiana benthamiana* 3–4 µg g^{-1}	Protection of 40% of mice immunized with crude extract containing E7 after tumour challenge	Franconi *et al.*, 2002
VPH-16 E7	PVX vector in *N. benthamiana* 15–20 µg g^{-1}	Protection of 80% of mice immunized with crude extract containing E7 after tumour challenge	Franconi *et al.*, 2006
L2 epitopes and HPV-16 E6	Capsid was used for potato virus A decorated with two epitopes, one of L2 (on the N terminus) and one of the E7 (at the C terminus) and using the vector PVX	ND	Cerovská *et al.*, 2008
VPH-16 E7GGG	DNA vaccine	Protection of immunized mice after challenge	Massa *et al.*, 2008
PVX CP– VPH-16 E7	Tobacco chloroplasts 0.5 to 0.1% of PTS	ND	Morgenfeld *et al.*, 2009
VPH-16 E7	PVX vector in *N. benthamiana*	Dendritic cells were treated with crude extract containing E7, inducing their maturation and in turn naïve lymphocytes induced them to produce a T-cell response specific for E7	Di Bonito *et al.*, 2009
L1 (HPV-16) with E6 and E7 epitopes	Transgenic tomatoes 0.05–0.1% of TSP	Protection of immunized mice after tumour challenge. Mice developed specific neutralizing antibodies and CTL	Paz De la Rosa *et al.*, 2009
VPH-16 E7GGG	PVX in *N. benthamiana*	ND	Plchova *et al.*, 2011
VPH-16 E7GGG	*N. benthamiana* 0.1 g protein kg^{-1} plant biomass	ND	Buyel *et al.*, 2012

ND, not determined.

challenge with C3 tumour cells. Antitumour activity of this preparation was enhanced by increasing the amount of E7 in the crude extract. Up to 80% of immunized mice did not present tumours after 50 days post-challenge. Interestingly, 100% of mice were protected by immunization with an *E. coli*-derived vaccine containing 10 µg of antigen (i.e. fourfold the dose employed with plants) in the presence of the QuilA adjuvant. This indicates that the plant crude extract possesses some adjuvant properties.

This activity and the possible use of E7 in cancer immunotherapy were investigated on dendritic cells (Di Bonito *et al.*, 2009). The extract was not toxic, it did not facilitate the entrance of the antigen to the cells, but it did induce maturation of the cells. Importantly, the extract was able to induce naïve lymphocytes to produce E7-specific T-cells. Application of this technology in humans will require characterization of the compounds present in the crude extract.

Since E7 is an oncogenic protein, it has been proposed that fusion of E7 to other proteins would reduce the oncogenic activity. Plants have been employed to this purpose. The mutated gene HPV-16 E7 (E7GGG), lacking the retinoblastoma binding site was expressed fused to the β-1,3-1,4 glucanase from *Clostridium thermocellum* acting as a carrier molecule in *N. benthamiana* plants. This fusion protein showed antitumour activity by inhibiting tumour growth and increased survival in C57BL/6 mice challenged with TC-1 tumour cells. Furthermore, therapeutic vaccination of 50 mice with this antigen consistently prevented tumour growth, even of established tumours (Venuti *et al.*, 2009). This same antigen was produced in a contained system using tobacco hairy root cultures, as this type of platform is more likely to comply with Good Manufacturing Practices accepted by the FDA (Skarjinskaia *et al.*, 2008).

The viral capsid of many plant viruses has been found to be highly immunogenic and that of PVX is remarkable in this sense as it is able to induce CD4+ specific responses. Therefore, E7GGG was fused to PVX capsid protein and was tested as DNA vaccine in mice (Massa *et al.*, 2008). The vaccine induced a meaningful immune cellular response, prevented the growth of TC-1-induced tumour cells and extended the survival of vaccinated mice in comparison with mice vaccinated with E7GGG alone.

E7 has also been fused to PVX capsid protein and expressed on chloroplasts of tobacco resulting in higher yields than the E7 alone (Morgenfeld *et al.*, 2009). This suggests that the PVX capsid protein is somehow able to stabilize the E7 protein in the stroma. As it is always important to bear in mind the oncogenic properties of E7, the E7 protein from HPV-8 with the binding site to retinoblastoma mutated (QGD) was expressed in *N. benthamiana* (Noris *et al.*, 2011). Crude extracts containing this construct induced seroconversion and specific cytotoxic response in mice.

An alternative that is proving very attractive is the development of HPV vaccines with prophylactic and therapeutic activities. To this end, the preparations should contain epitopes of the envelope proteins (L1 and L2) as well as from the E6 and E7 oncoproteins. Cerovská *et al.* (2012) expressed the potato virus A capsid protein including two epitopes, one from L2 (on the N terminus) and one from E7 (at the C terminus) in a PVX vector in transgenic *N. benthamiana* plants. The construct was found to be immunogenic in mice by two routes of administration. In another study, Paz De la Rosa *et al.* (2009) reported the expression in transgenic tomato plants of HPV-16 L1 fused to a string of E6 and E7 epitopes that mediated cytotoxic T-cell activity. The assembled VLPs were able to induce neutralizing antibodies and cytotoxic T cell activity against E6 and E7 epitopes and protected mice against a challenge with TC-1 tumour cells (Paz De la Rosa *et al.*, 2009). Plchova *et al.* (2011) engineered different versions of mutagenized E7 protein (E7GGG) and fused them to the coat protein of potato virus X on both the 5′ and 3′ ends and evaluated the influence of a linker connecting both peptides on their expression. Fusion proteins were successfully expressed in bacteria and plants and their reactivity and ability to form VLPs were evaluated with anti-E7 antibodies.

One limitation of HPV vaccines is that neutralizing antibodies induced by L1 are type-restricted (Wakabayashi *et al.*, 2002). Addition of other HPV types would be a viable approach but this implies a considerable increase in the overall cost of the vaccine. The L2 minor capsid protein has emerged as a strong candidate for the development of prophylactic HPV vaccines. L2 is immunogenically subdominant to L1 and a small segment is exposed on the surface of mature capsids. Epitope display on the surface of VLPs or capsomeres may improve the immunogenicity of L2 regions and broaden the protection of L1-based vaccines (Pineo *et al.*, 2013). Different epitopes and regions of the L2 protein alone or fused to the L1 protein have been expressed in bacteria and other systems but in plants there are very few reports. Palmer *et al.* (2006) fused L2 from rabbit HPV to the coat protein of the tobacco mosaic virus and expressed the fusion in tobacco plants. Administration of the vaccine to rabbits conferred complete protection against viral infection.

On the other hand, the N-terminus of L2 is highly conserved and can induce a broad range of protective cross-neutralizing antibodies *in vivo* (Pastrana *et al.*, 2005). This suggests that a monovalent vaccine could protect against a broad range of HPV types. However, the titres of neutralizing antibody against the HPV type from which the L2 vaccine was derived were higher than those against heterologous types and lower than those induced by L1 VLPs (Roden *et al.*, 2000; Pastrana *et al.*, 2005). Concatenation of L2 peptides from different viral types has been regarded as a possible solution to this problem and it has been a strategy tested recently with great success. These constructs, expressed in *E. coli*, were able to induce high neutralizing antibody titres which confer cross-protection against highly divergent HPV types (Jagu *et al.*, 2009, 2013). Pineo *et al.* (2013) expressed in *N. benthamiana* a HVP-16 L1 chimera, containing three cross-protective epitopes from L2 protein in exposed regions of the L1 protein. The chimeras were targeted to chloroplasts and were highly expressed with yields of ~1.2 g kg^{-1} plant tissue, however, they assembled differently, suggesting that the length and probably the nature of the L2 epitopes affect VLP assembly. Some chimaeras assembled into capsomeres and elicited weaker humoral immune responses (Pineo *et al.*, 2013).

HPV-16 L1 has also been employed as a carrier for epitopes from viruses other than HPV. Two epitopes from Influenza A virus M2e2-24 and M2e2-9 were fused to human codon optimized HPV-16 L1 and transiently expressed in plants using the cowpea mosaic virus-derived expression vector, pEAQ-HT (Matić *et al.*, 2011).

Stanley (2010) has suggested that besides the use of L2, L1 capsomeres as opposed to VLPs are promising second generation vaccines with a suitable adjuvant but, as mentioned before, capsomeres are not as immunogenic as VLPs.

Monroy-García *et al.* (2013) generated a novel HPV-16 L1-based chimeric virus-like particle produced in *N. benthamiana* containing a string of T-cell epitopes from HPV-16 E6 and E7 fused to its C-terminus and analysed the persistence of specific IgG antibodies with neutralizing activity induced by immunization with these cVLPs, as well as their therapeutic potential in a tumour model of C57BL/C mice. They observed that these chimeric particles induced persistent IgG antibodies for over 12 months, with reactivity and neutralizing activity for VLPs composed of only the HPV-16L1 protein. Efficient protection for long periods of time and inhibition of tumour growth induced by TC-1 tumour cells expressing HPV-16 E6/E7 oncoproteins, as well as significant tumour reduction (57%), were observed in mice immunized with these particles.

6.6 Conclusions and Perspectives

As has been described in this chapter, there is abundant evidence that HPV VLPs and antigens can be expressed in plants efficiently, in high yields and retaining their immunogenicity and efficacy in various animal models. Plants have proved their worth in the development of first and second generation HPV vaccines. The combination of prophylactic and therapeutic vaccines produced in plants has been investigated extensively yielding significant efficacy testing in animal models. Considering the complexity of the responses that a therapeutic vaccine may evoke, plants can be considered not only as a 'biofactory' producing biosimilars but also as a source of immunomodulators that can create the proinflammatory environment needed for elimination of the persistent HPV infection and related lesions.

Current strategies for the development of safe and effective prophylactic vaccines are based on the induction of neutralizing antibodies against the major (L1) and minor (L2) capsid proteins of HPV. To improve the existing strategies for broad protection, a conserved and cross-protective antigen such as L2 should be included in vaccine formulations. Ideally, second generation vaccines should broaden coverage, induce long-term protection at the mucosal level, be cheap and of easy administration and provide therapeutic as well as prophylactic efficacy (Giorgi *et al.*, 2010). If vaccination was extended to other segments of population (in many countries it is administered to very young people), HPV eradication would be accelerated as the data show in the USA (Markowitz *et al.*, 2013).

There are several second-generation HPV vaccines under investigation, including mixtures of more types of HPV from Merck, capsomere-based preparations, chimeric L1 proteins and vaccines including all or parts of the L2 minor capsid protein (Jagu *et al.*, 2009, 2013). Nevertheless, one important issue remaining is how the HPV vaccines will affect the prevalence of other HPV types if the vaccine strains are effectively eradicated.

The development and use of cheap plant-based vaccines against HPV finds its greatest justification in poor countries, where the highest incidence and infection of HPV occur. There are several obstacles that still need to be cleared (improvement of yields, clinical trials, governmental approval, etc.), but the expectations of delivering a cheap and potent HPV vaccine for poor countries remain high. However, similar to other biotechnological developments, most of the technology for plant-derived vaccines has been patented in rich countries. Poor countries, which usually have a high disease burden, often have poor or inexistent IP protection rules and lack of adequate knowledge and infrastructure to protect and commercialize a biotechnological product. There are many publications and grant proposals on plant-derived vaccines where poor countries are one of the justifications for the work and an emphasis is made on the necessity, some authors have called it a 'moral imperative' (Ma *et al.*, 2005), to provide low-cost medicines and vaccines to poor countries. It has even been suggested that plant-derived vaccines may be approved in an industrialized country and then be more broadly used in poor countries (Thanavala *et al.*, 2006; Waheed *et al.*, 2012).

There is a clear need to develop plant-based low-cost medicines and vaccines for poor countries. This technology may offer a new model, which may allow a wider participation, beyond the well-established multinational pharmaceutical companies. Poor countries would potentially be involved, although it is still not well defined how, and the focus could be on specific regional diseases that do not feature in current drug development programmes. It is hoped that this technology will eventually help those who need it the most and that the issue of IP does not represent an insurmountable obstacle. Putting the collective benefit ahead of the personal gain will be the key for the full realization of this technology.

Acknowledgement

JFCE is indebted to CONACYT for a MSc scholarship.

References

Biemelt, S., Sonnewald, U., Galmbacher, P., Willmitzer, L. and Müller, M. (2003) Production of human papillomavirus type 16 virus-like particles in transgenic plants. *Journal of Virology* 77, 9211–9220.

Bishop, B., Dasgupta, J. and Chen, X.S. (2007) Structure-based engineering of papillomavirus major capsid L1: controlling particle assembly. *Virology Journal* 4, 3.

Bosch, F.X., Lorincz, A., Muñoz, N., Mejier, C.J. and Shah, K.V. (2002) The causal relation between human papillomavirus and cervical cancer. *Journal of Clinical Pathology* 55, 244–265.

Bruni, L., Diaz, M., Castellsagué, X., Ferrer, E., Bosch, F.X. and de Sanjosé, S. (2010) Cervical human papillomavirus prevalence in 5 continents: meta-analysis of 1 million women with normal cytological findings. *Journal of Infectious Diseases* 202, 1789–1799.

Buyel, J.F., Bautista, J.A., Fischer, R. and Yusibov, V.M. (2012) Extraction, purification and characterization of the plant-produced HPV16 subunit vaccine candidate E7 GGG. *Journal of Chromatography B* 880, 19–26.

Cerovská, N., Hoffmeisterová, H., Pecenková, T., Moravec, T., Synková, H., Plchová, H. and Velemínský, J (2008) Transient expression of HPV16 E7 peptide (aa 44-60) and HPV16 L2 peptide (aa 108-120) on chimeric potyvirus-like particles using Potato virus X-based vector. *Protein Expression and Purification* 58, 154–161.

Cerovská, N., Hoffmeisterova, H., Moravec, T., Plchova, H., Folwarczna, J., Synkova, H., Ryslava, H., Ludvikova, V. and Smahel, M. (2012) Transient expression of Human papillomavirus type 16 L2 epitope fused to N- and C terminus of coat protein of Potato virus X in plants. *Journal of Biosciences* 37, 125–133.

Chih, H.J., Lee, A.H., Colville, L., Binns, C.W. and Xu, D. (2013) A review of dietary prevention of human papillomavirus-related infection of the cervix and cervical intraepithelial neoplasia. *Nutrition and Cancer* 65, 317–328.

Di Bonito, P., Grasso, F., Mangino, G., Massa, S., Illiano, E., Franconi, R., Fanales-Belasio, E., Falchi, M., Affabris, E. and Giorgi, C. (2009) Immunomodulatory activity of a plant extract containing human papillomavirus 16-E7 protein in human monocyte-derived dendritic cells. *International Journal of Immunopathology and Pharmacology* 22, 967–978.

Fernández-San Millán, A., Ortigosa, S.M., Hervás-Stubbs, S., Corral-Martínez, P., Seguí-Simarro, J.M., Gaétan, J., Coursaget, P. and Veramendi, J. (2008) Human papillomavirus L1 protein expressed in tobacco chloroplasts self-assembles into virus-like particles that are highly immunogenic. *Plant Biotechnology Journal* 6, 427–441.

Franconi, R., Di Bonito, P., Dibello, F., Accardi, L., Muller, A., Cirilli, A., Simeone, P., Donà, M.G., Venuti, A. and Giorgi, C. (2002) Plant-derived human papillomavirus 16 E7 oncoprotein induces immune response and specific tumor protection. *Cancer Research* 62, 3654–3658.

Franconi, R., Massa, S., Illiano, E., Mullar, A., Cirilli, A., Accardi, L., Di Bonito, P., Giorgi, C. and Venuti, A. (2006) Exploiting the plant secretory pathway to improve the anticancer activity of a plant-derived HPV16 E7 vaccine. *International Journal of Immunopathology and Pharmacology* 19, 187–197.

Frenzel, A., Hust, M. and Schirrmann, T. (2013) Expression of recombinant antibodies. *Frontiers in Immunology* 4, 217.

Giorgi, C., Franconi, R. and Rybicki, E.P. (2010) Human papillomavirus vaccines in plants. *Expert Review of Vaccines* 9, 913–924.

Gómez Lim, M.A. (2011) Plants as platform for production of pharmaceutical compounds. In: Liong, M.T. (ed.) *Bioprocess Sciences and Technology*. Nova Science Publishers, Hauppauge, New York, pp. 1–26.

Hongli, L., Xukui, L., Ting, L., Wensheng, L., Lusheng, S. and Jin, Z. (2013) Transgenic tobacco expressed HPV16-L1 and LT-B combined immunization induces strong mucosal and systemic immune response in mice. *Human Vaccines and Immunotherapeutics* 9, 83–89.

Jagu, S., Karanam, B., Gambhira, R., Chivukula, S.V., Chaganti, R.J., Lowy, D.R., Schiller, J.T. and Roden, R.B. (2009) Concatenated multitype L2 fusion proteins as candidate prophylactic pan-human papillomavirus vaccines. *Journal of the National Cancer Institutes* 101, 782–792.

Jagu, S., Kwak, K., Karanam, B., Huh, W.K., Damotharan, V., Chivukula, S.V. and Roden, R.B. (2013) Optimization of multimeric human papillomavirus L2 vaccines. *PLoS ONE* 8,1.

Kirnbauer, R., Booy, F., Cheng, N., Lowy, D.R. and Schiller, J.T. (1992) Papillomavirus L1 major capsid protein self-assembles into virus-like particles that are highly immunogenic. *Proceedings of the National Academy of Sciences USA* 89, 12180–12184.

Kohl, T., Hitzeroth, I.I., Stewart, D., Varsani, A., Govan, V.A., Christensen, N.D., Williamson, A.L. and Rybicki, E.P. (2006) Plant-produced cottontail rabbit papillomavirus L1 protein protects against tumor challenge: a proof-of-concept study. *Clinical and Vaccine Immunology* 13, 845–853.

Kohl, T.O., Hitzeroth, I.I., Christensen, N.D. and Rybicki, E.P. (2007) Expression of HPV-11 L1 protein in transgenic *Arabidopsis thaliana* and *Nicotiana tabacum*. *BMC Biotechnology* 7, 56.

Lenzi, P., Scotti, N., Alagna, F., Tornesello, M.L., Pompa, A., Vitale, A., De Stradis, A., Monti, L., Grillo, S., Buonaguro, F.M., Maliga, P. and Cardi, T. (2008) Translational fusion of chloroplast-expressed human papillomavirus type 16 L1 capsid protein enhances antigen accumulation in transplastomic tobacco. *Transgenic Research* 17, 1091–1102.

Liu, H.L., Li, W.S., Lei, T., Zheng, J., Zhang, Z., Yan, X.F., Wang, Z.Z., Wang, Y.L. and Si, L.S. (2005) Expression of human papillomavirus type 16 L1 protein in transgenic tobacco plants. *Acta Biochimica et Biophysica Sinica* 37, 153–158.

Love, A.J., Chapman, S.N., Matic, S., Noris, E., Lomonossoff, G.P. and Taliansky, M. (2012) In planta production of a candidate vaccine against bovine papillomavirus type 1. *Planta* 236, 1305–1313.

Ma, J.K.C., Barros, E., Bock, R., Christou, P., Dale, P.J., Dix, P.J., Fischer, R., Irwin, J., Mahoney, R., Pezzotti, M., Schillberg, S., Sparrow, P., Stoger, E. and Twyman, R.M. (2005) Molecular farming for new drugs and vaccines. *EMBO Reports* 6, 593–599.

Maclean, J., Koekemoer, M., Olivier, A.J., Stewart, D., Hitzeroth, I.I., Rademacher, T., Fischer, R., Williamson, A.L. and Rybicki, E.P. (2007) Optimization of human papillomavirus type 16 (HPV-16) L1 expression in plants: comparison of the suitability of different HPV-16 L1 gene variants and different cell-compartment localization. *Journal of General Virology* 88, 1460–1469.

Markowitz, L.E., Hariri, S., Lin, C., Dunne, E.F., Steinau, M., McQuillan, G. and Unger, E.R. (2013) Reduction in human papillomavirus (HPV) prevalence among young women following HPV vaccine introduction in the United States, National Health and Nutrition Examination Surveys, 2003-2010. *Journal of Infectious Diseases* 208, 385–393.

Massa, S., Simeone, P., Muller, A., Benvenuto, E., Venuti, A. and Franconi, R. (2008) Antitumor activity of DNA vaccines based on the human papillomavirus-16 E7 protein genetically fused to a plant virus coat protein. *Human Gene Therapy* 19, 354–364.

Matić, S., Rinaldi, R., Masenga, V. and Noris, E. (2011) Efficient production of chimeric human papillomavirus 16 L1 protein bearing the M2e influenza epitope in *Nicotiana benthamiana* plants. *BMC Biotechnology* 11, 106.

Matić, S., Masenga, V., Poli, A., Rinaldi, R., Milne, R.G., Vecchiati, M. and Noris, E. (2012) Comparative analysis of recombinant human papillomavirus 8 L1 production in plants by a variety of expression systems and purification methods. *Plant Biotechnology Journal* 10, 410–421.

Monk, B.J. and Tewari, K.S. (2007) The spectrum and clinical sequelae of human papillomavirus infection. *Gynecologic Oncology* 107, S6–S13.

Monroy-García, A., Gómez, L.M.A., Weiss, S.B., Hernández, M.J., Huerta, Y.S., Rangel, S.J.F., Santiago, O.E. and Mora, G.M. (2013) Immunization with an HPV-16 L1-based chimeric virus-like particle containing HPV-16 E6 and E7 epitopes elicits long-lasting prophylactic and therapeutic efficacy in an HPV-16 tumour mice model. *Archives of Virology* DOI:10.1007/s00705-013-1819-z.

Morgenfeld, M., Segretin, M.E. and Wirth, S. (2009) Potato virus X coat protein fusion to human papillomavirus 16 E7 oncoprotein enhance antigen stability and accumulation in tobacco chloroplast. *Molecular Biotechnology* 43, 243–249.

Münger, K., Baldwin, A., Edwards, K.M., Hayakawa, H., Nguyen, C.L., Owens, M., Grace, M. and Huh, K. (2004) Mechanisms of human papillomavirus induced oncogenesis. *Journal of Virology* 78, 11451–11460.

Noris, E., Poli, A., Cojoca, R., Rittà, M., Cavallo, F., Vaglio, S., Matic, S. and Landolfo, S. (2011) A human papillomavirus 8 E7 protein produced in plants is able to trigger the mouse immune system and delay the development of skin lesions. *Archives of Virology* 156, 587–595.

Palmer, K.E., Benko, A., Doucette, S.A., Cameron, T.I., Foster, T., Hanley, K.M., McCormick, A.A., McCulloch, M., Pogue, G.P., Smith, M.L. and Christensen, N.D. (2006) Protection of rabbits against cutaneous papillomavirus infection using recombinant tobacco mosaic virus containing L2 capsid epitopes. *Vaccine* 24, 5516–5525.

Pastrana, D.V., Gambhira, R., Buck, C.B., Pang, Y.Y., Thompson, C.D., Culp, T.D., Christensen, N.D., Lowy, D.R., Schiller, J.T. and Roden, R.B. (2005) Crossneutralization of cutaneous and mucosal Papillomavirus types with anti-sera to the amino terminus of L2. *Virology* 337, 365–372.

Paz De la Rosa, G., Monroy-García, A., Mora-García, M. de L., Peña, C.G., Hernández-Montes, J., Weiss-Steider, B. and Gómez-Lim, M.A. (2009) An HPV 16 L1-based chimeric human papillomavirus-like particles containing a string of epitopes produced in plants is able to elicit humoral and cytotoxic T-cell activity in mice. *Virology Journal* 6, 2.

Pineo, C.B., Hitzeroth, I.I. and Rybicki, E.P. (2013) Immunogenic assessment of plant-produced human papillomavirus type 16 L1/L2 chimaeras. *Plant Biotechnology Journal* 11, 964–975.

Plchova, H., Moravec, T., Hoffmeisterova, H., Folwarczna, J. and Cerovska, N. (2011) Expression of Human papillomavirus 16 E7ggg oncoprotein on N- and C-terminus of Potato virus X coat protein in bacterial and plant cells. *Protein Expression and Purification* 77, 146–152.

Regnard, G.L., Halley-Stott, R.P., Tanzer, F.L., Hitzeroth, I.I. and Rybicki, E.P. (2010) High level protein expression in plants through the use of a novel autonomously replicating geminivirus shuttle vector. *Plant Biotechnology Journal* 8, 38–46.

Roden, R.B., Yutzy, W.I., Fallon, R., Inglis, S., Lowy, D.R. and Schiller, J.T. (2000) Minor capsid protein of human genital papillomaviruses contains subdominant, crossneutralizing epitope. *Virology* 270, 254–257.

Scotti, N. and Rybicki, E.P. (2013) Virus-like particles produced in plants as potential vaccines. *Expert Review of Vaccines* 12, 211–224.

Skarjinskaia, M., Karl, J., Araujo, A., Ruby, K., Rabindran, S., Streatfield, S.J. and Yusibov, V. (2008) Production of recombinant proteins in clonal root cultures using episomal expression vectors. *Biotechnology and Bioengineering* 100, 814–819.

Sohn, U., Nam, H.G., Park, D.H. and Kim, K.H. (2002) Recombinant human papillomavirus vaccine expressed in transgenic plants. United States Patent 6444805.

Stanley, M. (2010) Prospects for new human papillomavirus vaccines. *Current Opininon in Infectious Diseases* 23, 70–75.

Thanavala, Y., Huang, Z. and Mason, H.S. (2006) Plant-derived vaccines: a look back at the highlights and a view to the challenges on the road ahead. *Expert Review of Vaccines* 5, 249–260.

Varsani, A., Williamson, A.L., de Villiers, D., Becker, I., Christensen, N.D. and Rybicki, E.P. (2003a) Chimeric human papillomavirus type 16 (HPV-16) L1 particles presenting the common neutralizing epitope for the L2 minor capsid protein of HPV-6 and HPV-16. *Journal of Virology* 77, 8386–8393.

Varsani, A., Williamson, A.L., Rose, R.C., Jaffer, M. and Rybicki, E.P. (2003b) Expression of human papillomavirus type 16 major capsid protein in transgenic *Nicotiana tabacum* cv. Xanthi. *Archives of Virology* 148, 1771–1786.

Varsani, A., Williamson, A.L., Stewart, D. and Rybicky, E.P. (2006) Transient expression of human papillomavirus type16 L1 protein in *Nicotiana benthamiana* using an infectious tobamovirus vector. *Virus Research* 120, 91–96.

Venuti, A., Massa, S., Mett, V., Vedova, L.D., Paolini, F., Franconi, R. and Yusibov, V. (2009) An E7-based therapeutic vaccine protects mice against HPV16 associated cancer. *Vaccine* 27, 3395–3397.

Waheed, M.T., Thönes, N., Müller, M., Hassan, S.W., Razavi, N.M., Lössl, E., Kaul, H.P. and Lössl, A.G. (2011a) Transplastomic expression of a modified human papillomavirus L1 protein leading to the assembly of capsomeres in tobacco: a step towards cost-effective second-generation vaccines. *Transgenic Research* 20, 271–282.

Waheed, M.T., Thönes, N., Müller, M., Hassan, S.W., Gottschamel, J., Lössl, E., Kaul, H.P. and Lössl, A.G. (2011b) Plastid expression of a double-pentameric vaccine candidate containing human papillomavirus-16 L1 antigen fused with LTB as adjuvant: transplastomic plants show pleiotropic phenotypes. *Plant Biotechnology Journal* 9, 651–660.

Waheed, M.T., Gottschamel, J., Hassan, S.W. and Lössl, A.G. (2012) Plant-derived vaccines: an approach for affordable vaccines against cervical cancer. *Human Vaccines and Immunotherapeutics* 8, 1–4.

Wakabayashi, M.T., Da Silva, D.M., Potkul, R.K. and Kast, W.M. (2002) Comparison of human papillomavirus type 16 L1 chimeric virus-like particles versus L1/L2 chimeric virus-like particles in tumor prevention. *Intervirology* 45, 300–307.

Warzecha, H., Mason, H.S., Lane, C., Tryggvesson, A., Rybicki, E., Williamson, A.L., Clements, J.D. and Rose, R.C. (2003) Oral immunogenicity of human papillomavirus-like particles expressed in potato. *Journal of Virology* 77, 8702–8711.

Zhou, J., Stenzel, D.J., Sun, X.Y. and Frazer, I.H. (1993) Synthesis and assembly of infectious bovine papillomavirus particles *in vitro*. *Journal of General Virology* 74, 763–768.

Zur Hausen, H. (2000) Papillomaviruses causing cancer: evasion host-cell control in early events in carcinogenesis. *Journal of the National Cancer Institutes* 92, 690–698.

7 Patenting of Plant-made Recombinant Pharmaceuticals and Access in the Developing World

Pascal M.W. Drake,* Sonia Sadone and Harry Thangaraj

Hotung Molecular Immunology Unit, Institute for Infection and Immunity, St George's University of London, UK

7.1 Disease and the Developing World

It has been estimated that more than 1400 species of infectious agents can cause disease in humans, of which approximately 350 are considered to be significant enough to warrant the gathering of information relevant to their diagnosis, epidemiology and therapy (Hay *et al.*, 2013). Worldwide, in 2002, infectious and parasitic diseases were responsible for 19% of deaths, predominantly from lower respiratory infections, HIV/AIDS, diarrhoeal disease, tuberculosis and malaria (http://www.who.int/whr/2004/annex/topic/en/annex_2_en.pdf). In addition to mortality, measurement of disability-adjusted life years (DALYs) indicates that infectious diseases account for approximately 25% of global morbidity (Brownlie *et al.*, 2006).

The burden of infectious diseases is much greater in the developing world, however non-communicable diseases (NCDs) are also responsible for a large and growing number of deaths in low- and middle-income countries (LMICs: we use the term interchangeably with 'developing world countries' and 'developing countries'). In all regions in 2004, except for Africa, over 50% of deaths in LMICs were due to NCDs (Mathers *et al.*, 2009). According to the World Health Organization, in 2004, 80% of cardiovascular disease-related deaths occurred in LMICs, and of the 170 million people with diabetes, two-thirds lived in the developing world (Marshall, 2004). The proportion of DALYs attributable to NCDs in developing world countries is estimated to increase from 33% in 2002 to 45% in 2030, due to a number of interrelated factors including population ageing, obesity, reduction in physical activity, changing diets and increased alcohol and tobacco use (Mathers *et al.*, 2009; http://www.who.int/mediacentre/news/releases/2011/ncds_20110427/en/index.html). Moreover, the interaction of infectious diseases and NCDs is another factor in the prevalence of disease in developing countries. For example, the risk of developing tuberculosis

---

* Corresponding author, E-mail: pdrake@sgul.ac.uk

(TB) is greater in diabetics (TB may in turn increase the risk of developing type 2 diabetes), and HIV increases the risk for human papilloma virus-induced cervical cancer threefold (Bygbjerg, 2012).

7.2 Plant-made Recombinant Pharmaceuticals

Historically, plants have been an extremely important source of medicines. From 1959 to 1973, 25% of all prescriptions dispensed from community pharmacies in the USA contained plant products (Farnsworth and Loub, 1983). More recently, plants have been investigated as production systems for recombinant protein pharmaceuticals, a process known as molecular farming. In 1982, human insulin from the genetically modified bacterium *Escherichia coli* became the first recombinant protein pharmaceutical to be licensed for use in humans. Approximately 150 recombinant pharmaceuticals, expressed mainly in microbial or mammalian cells, are now approved by either the Food and Drug Administration (FDA) or the European Medicines Agency (EMEA) for human clinical use (Ferrer-Miralles *et al.*, 2009). The first plant-made recombinant pharmaceutical (PMP), human growth hormone, was demonstrated in 1986 (Barta *et al.*, 1986) and since then, antibodies, blood products, enzymes, microbicides, hormones, structural protein polymers and subunit vaccine candidates have all been expressed in plants (Twyman *et al.*, 2005). A number of molecular farming candidates have entered clinical trials and the first regulatory approval for the use of a PMP in humans was obtained in 2012 for taliglucerase alfa (Elelyso™), an enzyme used in the treatment of Gaucher's disease (http://www.elelyso.com).

Transgenic plants offer several advantages over microbial and mammalian cell production systems for recombinant medicines. Molecular farming is an inexpensive production platform with low initial capital and maintenance costs, since plants are relatively inexpensive to establish and propagate. The farming of plants over large areas can also provide cost savings through economies of scale. Agricultural-scale growth of plants paves the way for many new treatments that require administration of vast amounts of PMP, such as application of a microbicide to mucosal surfaces, or production of a vaccine against pandemic disease. In contrast to microbes such as *E.coli*, plants, as eukaryotic organisms with appropriate chaperone and enzyme pathways, can produce complex multimeric proteins with appropriate post-translational modifications. This attribute was exemplified in 1995 with the expression in transgenic tobacco of fully assembled and functional secretory antibody, comprising ten polypeptide chains (Ma *et al.*, 1995).

Molecular farming is unique amongst the production systems in that it offers the possibility of multiple strategies for protein expression, including stable genetic transformation of either the nuclear or (prokaryotic) chloroplast genome, transient expression with deconstructed viral vectors, harvesting of PMP from secretions, and extraction from medium or individual cells grown in biofermentors. Further options are provided by the possible use of different plant species, diverse gene promoters (constitutive, inducible or organ specific) and intra-cellular targeting sequences. With this exceptional diversity, the production of any PMP, for any application, can be envisaged (Melnik and Stoger, 2013).

Molecular farming is a new technology, and although there has been a lack of significant commercial enablement to date, the licensing of Elelyso™ for human use may well lead to the approval of many other PMPs. We (Drake and Thangaraj, 2010) and others (Sparrow *et al.*, 2007) have contended that PMP production is a technology that is particularly suited to developing countries. As previously stated, molecular farming is relatively inexpensive, and much of the production process uses low-tech agricultural and food processing techniques, which could readily be implemented in developing countries. Indeed, many developing countries already have well-established procedures and facilities for horticultural production: for example Kenya is the third largest exporter of cut flowers in the world and in 2011 earned US$1.1 billion from the sale of flowers, fruits, vegetable and nut products (http://kenya.usaid.gov/programs/economic-growth/1253).

As an example of a potential application, a partially processed PMP for mucosal application could be produced locally in a developing country, transported and delivered without the need for refrigeration or specially trained medical practitioners, as administration by needle would not be required. PMP production could also be a valuable source of revenue for a developing country. Given the high retail value of many protein pharmaceuticals – treatment with the monoclonal antibody Soliris™ marketed by Alexion Pharmaceuticals for example costs US$400,000/year per patient – the potential for profit is obvious (http://www.pharmatimes. com/Article/13-01-25/High_cost_Soliris_not_backed_by_ministers. aspx).

7.3 Patenting of PMP

A patent, when granted, gives inventors exclusive rights on their invention for a limited period of time, usually 20 years. In return, the inventor is required to publish details of the invention as a pre-condition for obtaining the patent (Stott and Valentine, 2004). In the case of most authorities, the patent application is available for general inspection 18 months after the date of filing (Dunwell, 2006) and permission to use the patented invention can be granted through exclusive or non-exclusive (multiple licensee) licensing. In addition to the commercial sector, patenting activity is also undertaken by academic institutions, a process facilitated by the Bayh-Dole Act (Thursby and Thursby, 2003) in the USA and similar legislation elsewhere (Thangaraj *et al.*, 2009). Academic licensing of patents has led to the receipt of significant income for the sector: for example revenues of over US$1 billion were generated by US institutions in 2006 (Sampat, 2009).

Numerous patents have been issued throughout the development of transgenic plant technology, covering areas such as tissue culture, transformation techniques, and the genes of interest and co-transferred selectable markers (Dunwell, 2005). A total of 3622 patents were returned in a 2006 search of a US application database using the terms 'transgenic plant' and 'method' (Dunwell, 2006).

We previously investigated patenting of PMP between 2002 and 2008. 'Patents' referred to both granted patents and patent applications. Patent claims encompassed a recombinant product that had a therapeutic or prophylactic impact on health for a human or veterinary application. The principal findings from this study were that there was a clear downward trend in the number of patents filed between 2002 and 2008 and a greater number of patents were filed by public sector institutions or inventors than by the private sector. The US-dominated patenting activity providing nearly 30% of inventors. The majority of patents were for vaccine candidates (55%), followed by therapeutics (38%) and antibodies (7%) (Drake and Thangaraj, 2010).

Due to the immaturity of the PMP field, we cannot describe developing world access issues for these medicines, however, in the next section we shall discuss the various factors, including patents, that influence availability of pharmaceuticals in the developing world.

7.4 Access to Pharmaceuticals in the Developing World

Article 25.1 of the Universal Declaration of Human Rights (1948) states that 'Everyone has the right to a standard of living adequate for the health of himself and of his family, including food, clothing, housing and medical care and necessary social services'. In addition, three of the eight Millennium Development Goals pertain to human health. Despite these worthy declarations and despite the disproportionate burden of infectious and NCDs borne by populations in the developing world, access to medications in LMICs is far from universal (Lage, 2011).

With the exception of the role of patents, a comprehensive analysis of the factors influencing access to pharmaceuticals is beyond the scope of this article, however some of these issues will be briefly discussed. First, diseases that affect the developing world may not be a priority for pharmaceutical companies, as medicines for these conditions are unlikely to be as profitable as those targeting chronic diseases present in high-income countries. For example, between 1980 and 2004 only 1% of the medications developed were for tropical diseases and TB (Lage, 2011). The manufacture of pharmaceuticals is an expensive and complex undertaking, limited to a small number of countries and beyond the capabilities of the vast majority of LMICs.

In the majority of LMICs, medicines provided free of charge through the public sector are not available, and the majority of people do not have health insurance (Cameron *et al.*, 2009). Consequently, medicines must be paid for 'out of pocket' at the time of illness, a significant or unaffordable financial burden, which may result in individuals forgoing treatment or not completing a course of medicines (Niens *et al.*, 2010). The problem is compounded by the fact that in LMICs, medication prices in the private sector are often high, sometimes up to 80 times the international reference price (Lage, 2011).

A number of proposals have been suggested to try to ensure that developing world countries have access to pharmaceuticals. One such initiative is the formation of Product Development Partnerships (PDPs) – not-for-profit organizations which bring together public, private, academic and philanthropic sectors for the promotion of the development and access of pharmaceutical products that have limited commercial value (Brooks *et al.*, 2010). A second initiative has been the concept of 'tiered pricing', in which medicines are sold in LMICs at lower prices than in developed world countries (Moon *et al.*, 2011).

7.5 Patents and Access to Pharmaceuticals in the Developing World

The principal justification for patents is that they prevent competitors from copying inventions, allowing prices of the invention to be maintained at a level sufficiently high to recoup the initial investment and make a profit. In the absence of patents, it is argued, the incentive to invent is greatly reduced. In addition, the patent system encourages inventors to disclose the details of their invention thus allowing the spread of technological innovation (Sterckx, 2004). The effect of patenting on innovation in the development of, and access to, medicines has been extensively debated over a period of many years, and no consensus has been reached. A significant rise in patenting has been accompanied by a concomitant drop in the number of new pharmaceuticals on the market (http://www.cbo.gov/ftpdocs/76xx/doc7615/10-02-DrugR-D.pdf), but the question remains as to the effect, if any, of the former on the latter.

The debate regarding the effect of patents on access to medicines in the developing world has greatly intensified following the adoption of the Trade-related Aspects of Intellectual Property Rights (TRIPS) agreement by the World Trade Organization (WTO), which came into effect in industrial countries in 1996 (Attaran, 2004; Sterckx, 2004; http://www.wto.org/english/tratop_e/trips_e/tripfq_e.htm). The TRIPS agreement, applicable to members of the WTO, significantly strengthened patent protection standards. In 2001, in response to growing concerns that patent rules might restrict access to affordable medicines for populations in developing countries, the WTO drew up the Doha Declaration, which exempted the least developing countries from complying with TRIPS on pharmaceutical patents until 2016 with a further extension of this date a possibility (WTO; http://www.wto.org/english/tratop_e/trips_e/tripfq_e.htm).

Importantly, TRIPS allowed, under certain conditions, the issuing of compulsory licences whereby a government instructs a supplier to produce the patented product or process without the patent owner's consent. In

TRIPS, compulsory licences had to be granted mainly to supply the domestic market: this was amended in the Doha Declaration so that countries unable to manufacture the pharmaceuticals could obtain cheaper copies elsewhere if necessary (http://www.wto.org/english/tratop_e/trips_e/public_health_faq_e.htm).

Proponents of the patenting system state that the development of new drugs is a high risk activity: only 1 of every 5000–10,000 candidates tested in preclinical trials is approved, the average cost of bringing a pharmaceutical to market is US$1.2 billion and only one in three drugs brought to market is profitable (Glover, 2007). Consequently, patent protection is critical to ensure appropriate investment in research and development for the production of new medicines. Patents may also be particularly important for Small and Medium Enterprises (SMEs) in order to attract the funding required to develop and commercialize an invention. The pharmaceutical company GlaxoSmithKline (GSK) has argued that, rather than patent protection, the real barrier to medicine access in the developing world is poverty, manifested in the inability to pay for even the cheapest medicines, lack of healthcare infrastructure such as clinics and hospitals, poor distribution networks and low numbers of trained healthcare workers. In support of this argument, GSK points out that 95% of the medicines on the WHO Essential Drugs list are not patent protected and yet the WHO states that 30% of people in developing countries do not have reliable access to these drugs (http://www.gsk.com/content/dam/gsk/globals/documents/pdf/GSK-on-IP-and-access-to-medicines-in-developing-countries.pdf).

The opposite view is that the market exclusivity conferred by patents leads to profits that are much greater than the combined costs of research, development and production. According to the opponents of the patent system, in high income countries the evidence suggests that existing patents increase the cost of medicines and that in LMICs patents not only increase their cost but also the difficulty in developing novel mechanisms for medicine delivery (Saha *et al.*, 2003; Attaran, 2004; Gold *et al.*, 2010). Of particular concern for LMICs is that patents may increase the price of medicines, or hinder access, by restricting the production of lower cost generic alternatives (Sampat, 2009). Furthermore, according to the 'anti' view, the patent system does not incentivize the production of medicines for diseases that offer a commercially unattractive return such as tropical diseases prevalent in the developing world. It has also been suggested that patents in the pharmaceutical sciences can be manipulated to interfere with the practice of medicine. For example, companies use legal means to extend the period of market exclusivity for inventions, a process known as 'patent evergreening' (Kesselheim, 2007).

The arguments rage on, and agreement on the issue of patents and innovation in, and access to, medicine is not likely to be reached in the near future. However, as we discuss below, the form of licensing of patent rights may be very important in providing access to medicines.

7.6 Socially Responsible Licensing

In a previous study, we determined that over 50% of PMP product patents filed between 2002 and 2008 were by researchers from academia (Drake and Thangaraj, 2010). We have updated these findings by searching titles and abstracts of the World Intellectual Property Organization (WIPO) PATENTSCOPE search tool (http://patentscope.wipo.int/search/en/search.jsf) between 2009 and 2013 using the terms 'plant' and 'vaccine' or 'antibody' or 'pharmaceutical' or 'treatment'. Each return was scrutinized individually to identify patents pertaining to the use of a molecular farming product to treat disease. Again, as with our previous findings, approximately 50% of patents were from the public sector.

This significant contribution of public sector PMP patenting may have important ramifications when considering the licensing of patents, and by extension, may be another indication that PMPs are particularly suitable for developing world applications. Public sector institutions differ from companies in several important respects. They usually have multiple sources of revenues, such as government funding for teaching, student fees, philanthropic donations and grant funding for research, and therefore the need to make a profit from their inventions may be less acute than in the commercial sector. In addition, universities are usually populated by young, often highly idealistic students, who have the will and freedom to act in a manner that would not be possible for employees in the commercial sector. For example, in a well-known case, a sustained campaign by students at Yale University and others, led to Bristol Myers-Squibb (BMS) not asserting patent rights in South Africa for an HIV antiretroviral that had originally been licensed to BMS by Yale (Stevens and Effort, 2008).

It can be argued that there is a particular moral imperative to ensure that the inventions made by public sector institutions, which may have been principally funded by the tax payer and which are not responsible to shareholders, should be available to the economically disadvantaged in the developing world.

These considerations have led to the concept of socially responsible licensing (SRL), in which academic institutions license their patents with attention to eventual pricing, technology flow, and meeting underserved needs. Typically, the granting of a patent licence will require that the licensee makes available sub-licences to manufacturers in the developing world free of charge, or on preferential terms. Alternatively, the licensee may make available the products on preferential terms to developing countries.

An early example is provided by the University of California at Berkeley, in which a Socially Responsible Licensing Program (SRLP) was established in 2003 (Mimura, 2006). The SRLP was originated by the development and licensing of a hand-held system for the diagnosis of Dengue fever and has several goals, including the promotion of widespread availability of healthcare and technologies in the developing world. In 2003, the university consolidated IP management under the umbrella of Intellectual Property and Industry Research Alliances (IPIRA), which changed the metrics by which success in technology transfer was measured. In the words of Carol Mimura, assistant vice chancellor for IPIRA:

> in the current organizational structure, the future grant of a royalty-free license is financially detrimental to the revenue bottom of the IP licensing office, but since that strategy stimulates net funding, gifts, relationships and recognition to the campus that far outweigh the licensing revenue forgone, the benefit can be tallied in the social-impact bottom line and also (in some cases) as research revenue in the licensing office's peer division (the Industry Alliances Office).
>
> Mimura (2006)

A number of academic institutions in the USA and Canada are committed to the idea of global social responsibility and global health access. In an initiative spearheaded by Stanford University (USA), a white paper in progress that has signatories of several institutions shows the way forwards in dealing with licensing and patenting issues that may impact global health (http://news-service.stanford.edu/news/2007/march7/gifs/whitepaper. pdf). The paper states 'There is an increased awareness that responsible licensing includes consideration of the needs of people in developing countries and members of other underserved populations'. In addition, the Association of University Technology Managers (AUTM) in the USA has endorsed principles of responsible licensing for global health (http://www. autm.net/Content/NavigationMenu/TechTransferTechnologyTransfer Resources1/GHI.htm) and has developed a toolkit of sample clauses to aid technology transfer officers when making licensing deals that may impact on global health. Finally, several experts in the field have produced a joint initiative on SRL for technology transfer offices (available at http://www. accesstopharmaceuticals.org/wp-content/uploads/2013/04/SRL-Guide. Print-Version.pdf).

7.7 Conclusions

The 'golden rice' story may provide some indications of the role of patenting in the transfer of PMP technology to the developing world. Vitamin A deficiency, common in the developing world, is a major cause of childhood blindness, anaemia and reduced host resistance to infectious diseases (http://whqlibdoc.who.int/publications/2009/9789241598019_eng.pdf). Ingo Potrykus and colleagues in the Swiss Federal Institute of Technology developed golden rice, which was engineered to express the vitamin A precursor beta-carotene, initially at rather low levels although this was increased considerably in Golden Rice 2 (Ye *et al.*, 2000; Paine *et al.*, 2005). From the outset, Potrykus and colleagues were determined that the technology should be freely available to small-scale farmers. In addition, it was critical that public research institutions in developing countries had 'freedom to operate' (FTO) to introduce the trait into local varieties. Analysis of the intellectual property rights involved in the experiments undertaken in the golden rice project indicated that there were over 70 patents involved belonging to 32 different companies and universities, although the situation was somewhat simplified as only 12 of the patents related to developing countries (Potrykus, 2001; http://www.goldenrice.org). FTO was established, as all parties, including Monsanto, agreed to royalty-free licences applicable to farmers earning under US$10,000 per year from golden rice, while Zeneca received a licence for commercial use for income above this figure (Potrykus, 2001). Significantly, Potrykus was initially minded to join the opposition to patenting, but on further reflection came to the conclusion that the protection afforded by the process had been critical in the development and dissemination of the technology:

> It makes more sense to fight for a sensible use of Intellectual Property Rights (IPR)/Technical Property Rights (TPR). Thanks to public pressure there is much goodwill in the leading companies to come to an agreement on the use of IPR/TPR for humanitarian use that does not interfere with commercial interests of the companies.
>
> Potrykus (2001)

We believe that patents are crucial to the production of medicines, however, as we have described, patent licensing policy may result in greater access to medical innovations whilst still providing considerable benefits, both financial and non-pecuniary, to the institution. We are hopeful that the dominance of the public sector in the patenting of PMP will lead to the dissemination of any products generated using this technology to be supplied to the developing world on preferential terms.

Acknowledgements

PMWD is supported by the Sir Joseph Hotung Endowment. HT is supported by European Union Framework 7 grant no. 241839 entitled 'Access to Pharmaceuticals'.

References

Attaran, A. (2004) How do patents and economic policies affect access to essential medicines in developing countries? *Health Affairs* 23, 155–166.

Barta, A., Sommergruber, K., Thompson, D., Hartmuth, K., Matzke, M.A. and Matzke, A.J.M. (1986) The expression of a nopaline synthase – human growth-hormone chimeric gene in transformed tobacco and sunflower callus-tissue. *Plant Molecular Biology* 6, 347–357.

Brooks, A.D., Wells, W.A., McLean, T.D., Khanna, R., Coghlan, R., Mertensloetter, T., Privor-Dumm, L.A., Krattiger, A. and Mahoney, R.T. (2010) Ensuring that developing countries have access to new healthcare products: the role of product development partnerships. *Innovation Strategy Today* 3, 1–5.

Brownlie, J., Peckham, C., Waage, J., Woodhouse, M., Lyall, C., Meagher, L., Tait, J., Baylis, M. and Nicoll, A. (2006) *Foresight. Infectious Diseases: Preparing for the Future. Future Threats.* Office of Science and Innovation, London.

Bygbjerg, I. (2012) Double burden of noncommunicable and infectious diseases in developing countries. *Science* 337, 1499–1501.

Cameron, A., Ewen, M., Ross-Degnan, D., Ball, D. and Laing, R. (2009) Medicine prices, availability, and affordability in 36 developing and middle-income countries: a secondary analysis. *The Lancet* 373, 240–249.

Drake, P.M. and Thangaraj, H. (2010) Molecular farming, patents and access to medicines. *Expert Review of Vaccines* 9, 811–819.

Dunwell, J.M. (2005) Review: intellectual property aspects of plant transformation. *Plant Biotechnology Journal* 3, 371–384.

Dunwell, J.M. (2006) Patents and transgenic plants. *Proceedings of the Vth International Symposium on in Vitro Culture and Horticulture Breeding* 1–2, 719–732.

Farnsworth, N.R. and Loub, W.D. (1983) Information gathering and data bases that are pertinent to the development of plant-derived drugs. In: *Plants: The Potentials for Extracting Protein, Medicines and Other Useful Chemicals*. Workshop Proceedings, OTA-BP-F-23, U.S. Congress Of Technology Assessment, Washington, DC, pp. 178–195.

Ferrer-Miralles, N., Domingo-Espin, J., Corchero, J.L., Vazquez, E. and Villaverde, A. (2009) Microbial factories for recombinant pharmaceuticals. *Microbial Cell Factories* 8, 17–24.

Glover, G.J. (2007) The influence of market exclusivity on drug availability and medical innovations. *Aaps Journal* 9, E312–E316.

Gold, E.R., Kaplan, W., Orbinski, J., Harland-Logan, S. and Marandi, S. (2010) Are patents impeding medical care and innovation? *PloS Medicine* 7, e1000208.

Hay, S.I., Battle, K.E., Pigott, D.M., Smith, D.L., Moyes, C.L., Bhatt, S., Brownstein, J.S., Collier, N., Myers, M.F., George, D.B. and Gething, P.W. (2013) Global mapping of infectious disease. *Philosophical Transactions of the Royal Society B, Biological Sciences* 368, 20120256.

Kesselheim, A.S. (2007) Intellectual property policy in the pharmaceutical sciences: the effect of inappropriate patents and market exclusivity extensions on the health care system. *Aaps Journal* 9, E306–E311.

Lage, A. (2011) Global pharmaceutical development and access: critical issues of ethics and equity. *MEDICC Review* 13, 16–22.

Ma, J.K.C., Hiatt, A., Hein, M., Vine, N.D., Wang, F., Stabila, P., Van dolleweerd, C., Mostov, K. and Lehner, T. (1995) Generation and assembly of secretory antibodies in plants. *Science* 268, 716–719.

Marshall, S.J. (2004) Developing countries face double burden of disease. *Bulletin of the World Health Organization* 82, 556.

Mathers, C.D., Boerma, T. and Fat, D.M. (2009) Global and regional causes of death. *British Medical Bulletin* 92, 7–32.

Melnik, S. and Stoger, E. (2013) Green factories for biopharmaceuticals. *Current Medicinal Chemistry* 20, 1038–1046.

Mimura, C. (2006) Technology licensing for the benefit of the developing world: UC Berkeley's Socially Responsible Licensing Program. *Journal of the Association of University Technology Managers* 18, 15–28.

Moon, S., Jambert, E., Childs, M. and von Schoen-Angerer, T. (2011) A win–win solution?: A critical analysis of tiered pricing to improve access to medicines in developing countries. *Globalization and Health* 7, 39–49.

Niens, L.M., Cameron, A., Van de Poel, E., Ewen, M., Brouwer, W.B. and Laing, R. (2010) Quantifying the impoverishing effects of purchasing medicines: a cross-country comparison of the affordability of medicines in the developing world. *PLoS Medicine* 7, e1000333

Paine, J.A., Shipton, C.A., Chaggar, S., Howells, R.M., Kennedy, M.J., Vernon, G., Wright, S.Y., Hinchliffe, E., Adams, J.L., Silverstone, A.L. and Drake, R. (2005) Improving the nutritional value of Golden Rice through increased pro-vitamin A content. *Nature Biotechnology* 23, 482–487.

Potrykus, I. (2001) Golden rice and beyond. *Plant Physiology* 125, 1157–1161.

Saha, A., Grabowski, H., Birnbaum, H.G., Bizan, O., Greenberg, P.E. and Whitney, S. (2003) Generic competition in the pharmaceutical industry. *Value in Health* 6, 207.

Sampat, B.N. (2009) Academic patents and access to medicines in developing countries. *American Journal of Public Health* 99, 9–17.

Sparrow, P.A.C., Irwin, J.A., Dale, P.J., Twyman, R.M. and Ma, J.K.C. (2007) Pharma-Planta: Road testing the developing regulatory guidelines for plant-made pharmaceuticals. *Transgenic Research* 16, 147–161.

Sterckx, S. (2004) Patents and access to drugs in developing countries: an ethical analysis. *Developing World Bioethics* 4, 58–75.

Stevens, A.J. and Effort, A.E. (2008) Using academic license agreements to promote global social responsibility. *Journal of the Licensing Executive Society International* XLIII, 85–101.

Stott, M. and Valentine, J. (2004) Gene patenting and medical research: a view from a pharmaceutical company. *Nature Reviews Drug Discovery* 3, 364–368.

Thangaraj, H., van Dolleweerd, C.J., McGowan, E.G. and Ma, J.K.C. (2009) Dynamics of global disclosure through patent and journal publications for biopharmaceutical products. *Nature Biotechnology* 27, 614–619.

Thursby, J.G. and Thursby, M.C. (2003) Intellectual property – University licensing and the Bayh-Dole Act. *Science* 301, 1052.

Twyman, R.M., Schillberg, S. and Fischer, R. (2005) Transgenic plants in the biopharmaceutical market. *Expert Opinion in Emerging Drugs* 10, 185–218.

Ye, X.D., Al-Babili, S., Kloti, A., Zhang, J., Lucca, P., Beyer, P. and Potrykus, I. (2000) Engineering the provitamin A (beta-carotene) biosynthetic pathway into (carotenoid-free) rice endosperm. *Science* 287, 303–305.

8 Case Study 1: Rabies

Elizabeth Loza-Rubio* and Edith Rojas-Anaya

National Center of Microbiology in Animal Health (CENID-Microbiología), INIFAP, Mexico

8.1 Introduction

Since antiquity, rabies has been one of the most feared diseases. Human rabies remains an important public health problem in many developing countries. The World Health Organization (WHO) reports that more than 55,000 people die of this disease every year (WHO, 2008). Most of these cases occur in the developing countries. In most countries of Latin America, the major reservoirs are the dog, and lately the haematophagous bat (*Desmodus rotundus*), which is present in tropical and subtropical areas from northern Mexico to northern Argentina and Chile and transmits the disease mainly to cattle (Loza-Rubio *et al.*, 2005; Delpietro *et al.*, 2009). Vampire-bat attacks on cattle are a major concern for cattle-raising areas. Blood loss and paralytic rabies due to bat bites can impose severe losses on the livestock industry (Arellano-Sota, 1988).

A tome which focuses upon some of the major rabies issues spanning the geographical extent from the USA/Mexico border to Tierra del Fuego is long overdue, not the least because Latin America is rich with historical, cultural, ecological and viral diversity. One can only speculate about the primordial state of this disease, before canine rabies was imported during the 16th century with European colonization.

Clearly, the region has the greatest known diversity of rabies virus variants associated with the Chiroptera (the evolutionary well spring of the genus *Lyssavirus*), with representatives of major hosts among at least three bat families. Additionally, since the beginning of the 20th century, a complex epizootiological relationship was identified between rabies viruses and haematophagous bats, leading to bovine paralytic rabies – unique in the entire globe. Similarly, only in the New World are non-human primates (e.g. marmosets in Brazil) believed to serve primary rabies virus reservoirs.

The *Lyssavirus* genus encompasses 14 viruses: rabies virus (RABV), Lagos bat virus (LBV), Mokola virus (MOKV), Duvenhage virus (DUVV), European bat lyssavirus 1 and 2 (EBLV 1 and 2), Australian bat lyssavirus, Aravan virus, Khujand virus, Irkut virus, West Caucasian bat virus (WCBV) and Shimoni bat virus (SHIBV) (Loza-Rubio *et al.*, 1996; Nadin-Davis *et al.*,

* Corresponding author, E-mail: loza.elizabeth@inifap.gob.mx

2002; Kuzmin *et al.*, 2010; Dietzgen *et al.*, 2011). Recently, two new species of *Lyssavirus* have been identified. One has been isolated from an insectivorous bat (*Myotis nattererii*) in Germany and identified as Bokeloh virus (Freuling *et al.*, 2011), and the other has been isolated from an African civet and identified as Ikoma virus (IKOV) (Marston *et al.*, 2012). Classification of the genus is presented in Table 8.1.

Although several types of *Lyssavirus* are recognized worldwide, currently in the Americas only Genotype 1 has been identified, even though there are several groups carrying out epidemiological surveillance in order to verify this situation or if at one point in time any other type can be identified (Loza-Rubio *et al.*, 2012a).

8.2 Virus Structure

The virus genome is constituted by single-strand negative polarity ribonucleic acid (RNA). This means that the RNA cannot directly modify protein synthesis and must therefore be copied into a positive polarity

Table 8.1. Classification, geographical distribution and affected species by Lyssavirus genus (Based on Food and Agriculture Organization of the United Nations (2011) Newman, S.H., Field, H.E., de Jong, C.E. and Epstein, J.H. (eds) *Investigating the Role of Bats in Emerging Zoonoses: Balancing Ecology, Conservation and Public Health Interests.* FAO, Rome.).

Virus	Acronym	Maintenance hosts	Geographical distribution
Rabies virus (RV)	VRAB	Carnivora and multiple species of insectivorous and haematophagous bats	Worldwide (except some islands)
Lagos bat virus (LBV)	LBV	Bats	Africa
Virus Mokola (MOKV)	MOKV	Humans, cats, dogs, rodents, shrew	Africa
Virus Duvenhage (DUVV)	DUVV	Insectivorous bat	Africa
European bat lyssavirus 1 (EBLV-1)	EBLV-1	Insectivorous bat (*Eptesicus pipistrellus*)	Europe
European bat lyssavirus 2 (EBLV-2)	EBLV-2	Insectivorous bat (*Myotis* spp.)	Europe
Australian bat lyssavirus (ABLV)	ABLV	Insectivorous and frugivorous bats (suborder: *Megachiroptera/ Microchiroptera*)	Australia
Aravan virus (ARAV)	ARAV	Insectivorous bat (*Myotis blythi*)	Asia central
Khujand virus (KHUV)	KHUV	Insectivorous bat (*Myotis mystacinus*)	Asia central
Irkut virus (IRKV)	IRKV	Insectivorous bat (*Murina leucogaster*)	East Siberia
West Caucasian bat virus (WCBV)	WCBV	Insectivorous bat (*Miniopterus schrreibersi*)	Caucasus region, central Asia
Shimoni	BBLV	Unconfirmed – single isolate from *Hipposideros commersoni* (Commerson's leaf-nosed bat)	Africa
Bokeloh		Insectivorous bat (*Myotis nattererii*)	Europe
Ikoma		Civet	Africa

mRNA chain, giving rise to messenger RNA. As a result, the virus must have its own RNA-dependent RNA polymerase.

The mean length of the infectious or mature viral particles is 180 nm with an average diameter of 75 nm. The surface of the viral particle is covered with 400 glycoprotein projections. One of the ends of the viral particles is flat and the other is rounded, thus it is said that it has a bullet form (Fig. 8.1) (Shors, 2009).

The envelope is formed by a lipid layer that has five proteins on its surface: G protein (glycoprotein), alternates with the M protein (matrix protein). The nucleocapsid is formed by the N (nucleoprotein), P (phosphoprotein) and L (polymerase) proteins (Fig. 8.1).

8.2.1 Glycoprotein

Glycoprotein (G) is a protein with 505 amino acids weighing 65 to 67 kDa (Ross *et al.*, 2008). This protein is used by the virus to join with the host cells and initiates the relationship between them when the cell receptors link. The three protein type membrane receptors for the rabies virus that have been identified are: (i) the nicotinic receptor for acetylcholine; (ii) the low neurotrophin affinity receptor; and (iii) the neural cell adhesion molecule (NCAM). These receptors are implicated in the adsorption of the rabies virus and the promotion of infection directly into the nerve ends and/or gangliosides located in neurons, or at the point of axoplasmic transport of muscles (Lafon, 2005). The virus moves along the dorsal ganglions and

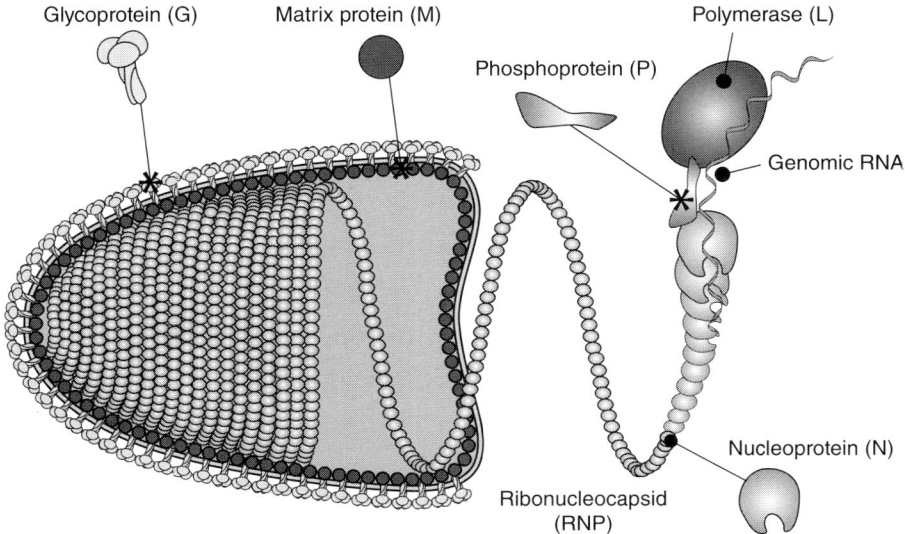

Fig. 8.1. Structure of rabies virus (http://viralzone.expasy.org/all by species/2.html). ©ViralZone 2008, Swiss Institute of Bioinformatics.

spinal cord; the brain is quickly infected causing apoptosis of the nerve cells and T cells (Lafon, 2011).

The G protein is also a target of T lymphocytes and induces the formation of neutralizing antibodies against the virus (Loza-Rubio *et al.*, 1998; Morales *et al.*, 2006). This is the reason why this protein has been used for making vaccines, since it is the most exposed antigen of the virus.

8.2.2 Matrix protein

The matrix protein (M) is a peripheral membrane protein with 200 amino acids weighing between 22 and 35 kDa. It participates in virus assembly and its exit from the cell, as well as in the down-regulation of transcription (Woldehiwet, 2002).

8.2.3 Nucleoprotein

The nucleoprotein (N) is constituted by 450 amino acids and weighs between 58 and 62 kDa. It intervenes in humoral and cellular immunity and it is the most important in terms of diagnosis due to its antigen role (Morales *et al.*, 2006) since it is the most conserved protein. It represents 90% of the nucleocapsid; it encapsulates and protects the genome from degrading. It has functional domains that join with the RNA, the P protein and possibly with the M protein.

8.2.4 L protein

This has a 190 kDa molecular weight. In association with the phosphoprotein it directs the transcription and replication of the viral RNA (Morales *et al.*, 2006; Jackson and Wunner, 2007).

8.2.5 Phosphoprotein

The phosphoprotein (P), which weighs between 35 and 40 kDa, interacts with the L protein by stabilizing it and participates in the encapsulation of the RNA. It interacts with the N protein preventing it from self-aggregation, and with the N-RNA complex it facilitates the initiation of RNA transcription and chain extension by RNA polymerase (L protein) (Romero *et al.*, 2006; Jackson and Wunner, 2007).

8.3 Vaccination Against Rabies – General Aspects

During the 1990s in Latin America the main reservoir of rabies to humans was dog. Thanks to the massive vaccination campaigns the levels of rabies have been reduced drastically. Nevertheless, currently rabies transmitted by haematophagous bats (*Desmodus rotundus*) represents a public health problem. This type of bat is widely distributed in tropical areas of America, from Mexico down to Argentina (Greenhall, 1993). It is important to consider that recently haematophagous bat populations have widened their area of action by occupying areas with higher altitude above sea level.

As mentioned previously one of the most effective strategies for the prevention, control and eradication of rabies is vaccination. Biologicals that have been used in humans are produced in nervous tissue and in cell culture. Together they represent a total of 50 million doses per year applied. In India, approximately 700,000 individuals require post-exposure treatment each year; it has been documented that every 2 s a dog bites a human and that every 30 min a person dies of rabies in that country.

8.4 New Generation Vaccines

As mentioned above, the G protein is critical for the host immune response to rabies virus infection because it is responsible for the induction of neutralizing antibodies and acts as a target for virus-specific helper and cytotoxic T cells (Johnson *et al.*, 2010). Therefore, this protein can be used as an immunogen when expressed in vectors such as vaccinia, canarypox, adenovirus, yeast, and in DNA vaccines, as well as in transgenic plants (Kieny *et al.*, 1984; Cadoz *et al.*, 1992; Xiang *et al.*, 1996; Sakamoto *et al.*, 1999; Henderson *et al.*, 2009; Tacket, 2009; Ventini *et al.*, 2010). Another protein used is the N protein, which can also prime T cells and induce rabies virus N-specific antibodies (Hooper *et al.*, 1994), and contains immuno-modulatory properties (Loza-Rubio *et al.*, 2009).

8.5 Plant-derived Vaccines

The literature indicates several potential advantages related to plant-derived vaccines, for example heat-stable formulation for storage and transport (avoiding cold chain), which is important in tropical and subtropical areas, eases delivery for better compliance leading to a reduced demand for skilled healthcare professionals in developing and developed countries.

The use of recombinant gene technologies by the vaccine industry has revolutionized the way antigens are generated and has provided safer, more effective means of protecting host organisms against bacterial, viral and parasitic pathogens (Lamphear *et al.*, 2002). In the case of viruses, no alternative to vaccines exists for animals since there are no antiviral drugs suitable for widespread application in the field. This underlines the need for controlling viral diseases of animals by vaccination. Advances in genetic engineering have made it possible to insert heterologous genes into several plant species, such as cereals and legumes. Plants are increasingly recognized as legitimate systems for the production of recombinant proteins and antigens. A wide range of proteins has been expressed and used for diagnostic purposes, industrial and pharmaceutical production of enzymes, food additives, therapeutic proteins, antibodies and vaccine antigens (Streatfield, 2006). However, despite nearly 20 years of development, there are only two plant-produced vaccine-related products that have gone all the way through all production and regulatory hurdles (Rybicki, 2009).

8.6 Expression of Rabies Virus Antigen in Plants

Oral vaccines that have been developed against the rabies virus have been successful in promoting protection antibodies. In fact, oral vaccination with Raboral V-RG vaccines (vaccinia recombinant virus vaccines that express the G protein) (Kieny et al., 1984; Fry et al., 2013) and with Rabigen SAG 2 (double mutant avirulent SAG2 strain) (Hanlon et al., 2002) has been effective in controlling rabies in the wild in Europe to such a degree that it has resulted in almost the total eradication of rabies in the wild in Western Europe (Brochier et al., 1996), and is successful in its control in Canada, the USA and other countries (Mainguy et al., 2013).

The G protein, which is the main antigen of the rabies virus, has also been expressed in tomato, tobacco and spinach plants. The first experience with the expression of the G protein of the rabies virus was in tomatoes (McGarvey et al., 1995). The full G gene of the virus was cloned into the BIN19 vector downstream of the 35S cauliflower mosaic virus (CaMV) promoter. Later, tomato cells were transformed by infection with *Agrobacterium tumefaciens*. The expressed glycoprotein was purified by immunoprecipitation from leaves and fruits, with two bands, one of 60 kDa and another of 62 kDa, detected with Western blot. This variation in protein weight could be due to differential glycosylation in the plant cell. The amount of recombinant glycoprotein in leaves was between approximately 1 and 10 ng mg^{-1} of soluble protein, while fruits had lower amounts.

Furthermore, other studies have reported that the oral administration of the rabies virus ribonucleoprotein induces the production of neutralizing antibodies when afterwards an inactive virus vaccine is applied in mice. As a way to improve the expression of proteins in plants, other viruses have been used as vectors to infect plant tissue such as the alfalfa mosaic virus (AIMV) using the coat protein (CP), which serves as a carrier for the peptides to be expressed. In this manner, Yusibov and collaborators (1997) expressed using this system the G and N proteins of the rabies virus and the human immunodeficiency virus type 1 (HIV-1). These constructs were inoculated into tobacco plants (*Nicotiana benthamiana*) in order to later isolate the virus from the leaves and semi-purify the viral particles for their inoculation of mice. Animals received seven doses (10 µg per dose) via intraperitoneal injection and assessing the response in the presence or absence of adjuvant. Using antibodies against the CP, the presence of a 28.9 kDa band was found in Western blot which corresponds to the fusion protein formed by the CP of AIMV and the viral peptides of the rabies virus. The identification of each peptide was carried out using monoclonal antibodies against each of them. Finally, it was demonstrated that the viral particles that were inoculated promoted an immune response in mice against the rabies virus antigens, as well as those of HIV-1, regardless of whether the adjuvant was present or not.

In another study, using the constructs reported by Yusibov, infection of tobacco and spinach plants was carried out (Modelska *et al.*, 1998). In this study, mice were immunized with transformed leaves by oral and intraperitoneal routes. Inoculation was carried out using 50 µg of purified recombinant virus in three doses. These same particles were administered orally through gastric intubation in four doses (250 µg per dose). Another group was fed for 7 days with the transformed spinach leaves (1 g per dose containing 15 µg of antigen). In all groups, serum samples and faecal pellets were collected 2 days before each immunization and the neutralizing activity of rabies virus-specific serum antibodies was determined. In animals immunized via the intraperitoneal route, the presence of antibodies was observed after the second immunization. The mice immunized by oral route showed the presence of IgG and IgA specific for both rabies viruses. The higher levels of immune response generated by the leaf-feeding approach as compared with gastric intubation raises the possibility that the plant cells enhanced the delivery of virus particles to the sites of immune responses. However, we do not exclude the possibility that the low dose of antigen given to mice during spinach feeding may have stimulated elevated IgA synthesis. A total of 40% of the animals survived the challenge.

Using this same transitory expression system as in the two previous studies, Yusibov and collaborators in 2002 assessed these plants not only in mice but also in people (Yusibov *et al.*, 2002). In the oral immunity study, three lots of spinach (3000 plants) were inoculated with the recombinant virus that expresses the peptides of the rabies virus. Mice were immunized via the intraperitoneal route with the purified recombinant protein (250 mg = 35 µg of peptide per dose) together with Freund's adjuvant and later challenged. Two groups were formed in the experiment with people. The first group (five individuals) were previously immunized against rabies and then given food using 20 g of fresh transformed spinach containing 0.6 mg of recombinant virus (84 µg of protein). The second group was composed of nine volunteers without previous immunization who received 150 g of fresh spinach tissue per dose (700 µg of protein).

Afterwards they received a dose of commercial vaccine intramuscularly and the presence of IgG and IgA was determined in serum. The leaves of spinach were found to contain 0.4 ± 0.007 mg of recombinant virus in fresh tissue that contained 84 mg of the chimeric peptide. A 19.3 kDa band, corresponding to the fusion peptide, was detected using Western blot. All (100%) of the mice immunized with the extract survived the challenge, 43% of those immunized with the synthetic peptides and 20% of those immunized with the alfalfa mosaic virus. Furthermore, three of the five volunteers mounted an effective response against the antigen after ingesting the transformed spinach. In six of the nine volunteers the antibody titres increased against the recombinant virus. Four of these individuals showed IgG and two out of seven showed IgA. In five out of nine of the volunteers there was an increase in IgG in serum after receiving three doses of the spinach leaves. Tolerance was not observed in any of these experiments.

Using tobacco, Ashraf and collaborators in 2005 (Ashraf *et al.*, 2005) expressed the glycoprotein fused to an endoplasmic reticulum retention sequence (SEKDEL) in order to improve its expression. The gene was cloned downstream of a double CaMV 35S promoter. Transformation was measured using *Agrobacterium tumefaciens*. The protein was purified and 25 μg of this extract was used to immunize five mice via the intraperitoneal route, receiving boosters at days 7, 14 and 28. These were later challenged using a standard laboratory strain (CVS). The protein purified from plant leaves showed a single band of ~66 kDa corresponding to G protein. This chimeric G-protein only represented 0.38% of the total soluble protein. The rabies glycoprotein expressed in tobacco is glycosylated and is not degraded during purification. These proteins show immunoreactivity to antirabies virus antibodies and elicit a high level of immune response in mice. The plant-derived G protein gave 100% protection similar to the commercial vaccine. In comparison with other studies, the protein that accumulates in tobacco is of higher molecular mass compared to the native protein.

In Mexico, edible vaccines have been developed using maize and carrots and have proved their efficiency in mice. In some cases these provided 100% protection in animals when challenged with a lethal virus originating from vampires (Loza-Rubio *et al.*, 2008; Rojas-Anaya *et al.*, 2009).

In the first report carried out by our study group, we reported the expression of the gene that codes for the glycoprotein in maize. Maize is a cereal that is rich in protein and is used for both human and animal consumption; this plant has been an adequate experimental model due to the high expression levels of transgenes that are obtained. Due to the above, it was perceived as a species with great potential to be used in the production of an edible vaccine. The vector pGHCNS was used for the transformation, cloning the complete G gene of the common rabies virus downstream of the maize ubiquitin and the CaMV 35S promoters; the expression cassette was flanked by a matrix attachment region (MARs). Maize embryogenic calluses were transformed with the above construction by biolistics (Fig. 8.2).

Regenerated maize plants were recovered and grown in the greenhouse (Fig. 8.3). The presence of the G gene and its products were detected in vegetal tissue by PCR and Western blot.

A fine powder was prepared from transformed grains and administered as a pellet (50 μg of recombinant G protein). Another group of mice was immunized intramuscularly with 50 μg of G protein using a commercial vaccine. All mice were challenged intracerebrally at day 90 post-vaccination using a vampire bat rabies virus that is used to evaluate commercial vaccine in Mexico. Embryogenic calli were transformed by biolistics and herbicide-resistant plants were obtained. Twenty-five plants were recovered and 92% contained the G-gene as detected by PCR. The rabies G protein was identified by Western blot and presented with a size of about 69 kDa. This increase in molecular weight may be due to post-translational modifications and this modification does not seem to have any adverse effect on the

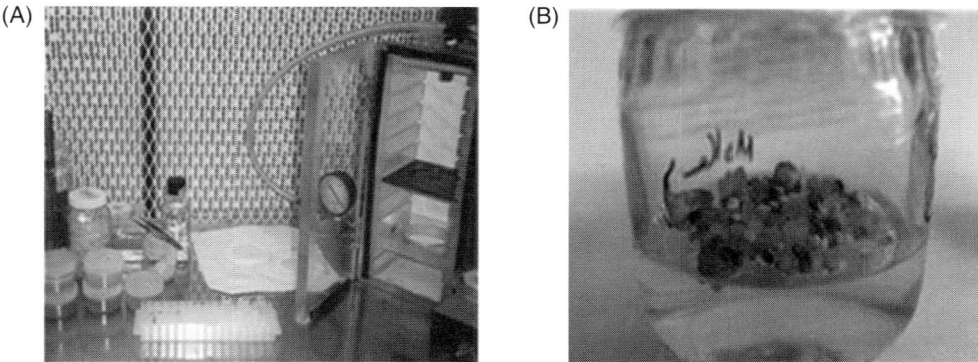

Fig. 8.2. Transformation procedure. (A) Transformation by biolistics of the maize embryogenic callus using the recombinant vector (photo courtesy of E. Loza-Rubio). (B) Selection of transformed maize embryogenic calli, using ammonium glufosinate (photo courtesy of E. Rojas-Anaya).

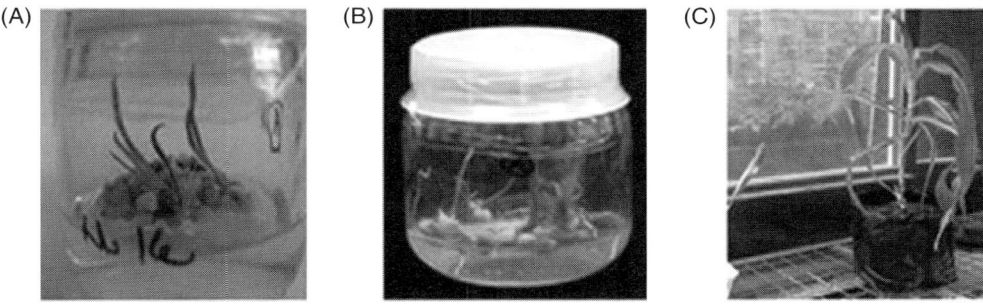

Fig. 8.3. Different growth stages of the edible vaccine against rabies: (A) embryogenic callus, (B) *in vitro* maize plantlets and (C) maize in greenhouse (photos courtesy of E. Loza-Rubio).

antigenic properties of the protein. Similar modifications were shown by McGarvey *et al.* (1995). Protein was expressed at 1% of total soluble protein, which is equivalent to about 50 µg g^{-1} fresh weight. This study obtained higher levels of expression than reported previously. The levels of expression obtained in this study are comparable to results obtained by others in maize when expressing the spike protein of the transmissible gastroenteritis coronavirus and the fusion protein of Newcastle disease virus. In sera, mice were seronegative at the start the experiment but by day 90 post-vaccination titres varied by more than 0.5 IU. The animals were protected at 100%, similar to the commercial vaccine. This work has demonstrated that the systems for transformation, selection and regeneration of maize developed in this study are efficient. Likewise, the plant-based G protein was able to induce viral neutralizing antibodies and protect mice after challenge.

As mentioned previously, another important antigen for the rabies virus is the nucleoprotein and is an alternative candidate antigen for development. Therefore, we set out to determine whether inoculation with a full-length, plant-based N protein expressed in tobacco and tomato conferred protection against rabies challenge in mice (Perea-Arango *et al.*, 2008). For this purpose, a full-length cDNA of the N gene was employed. The cDNA sequence was cloned at the SstI/XbaI restriction enzyme sites, so that the N sequence was under the control of the constitutively expressed 35S promoter and 35S terminator from CaMV. The resulting vector was designated pUCpss-Nrab. The expression cassette of pUCpss-Nrab was subcloned into pCambia 2301, creating the recombinant plasmid pCambia 2301-Nrab. The T-DNA regions also contained the neomycin phosphotransferase II gene (*nptII*) that provides kanamycin resistance. The pCambia 2301-Nrab plasmid DNA was introduced into *A. tumefaciens* strain LBA4404 by electroporation and subsequently employed for transformation of the *Lycopersicon esculentum* cultivar. For agroinfiltration, the N gene was released of plasmid pUCpss-Nrab and inserted into the vector pICH10990 creating the recombinant vector pICH10990-RabN. The constructions were transferred by electrophoration into *A. tumefaciens* to transform *N. benthamiana* by agroinfiltration. The presence of the N gene in regenerated plants was confirmed by PCR and Southern blot in plant tissue and the presence of the N protein was confirmed by Western blot using monoclonal anti-N antibody. The levels of N expressed in tomato fruit ranged from 0.1 to 0.5 mg g^{-1} and 1 to 5% total soluble protein (TSP), but in *N. benthamiana* leaves it reached levels of up to 4 mg g^{-1} and 45% TSP. The immune response was evaluated in mice immunized with plant extract. For tomatoes, group 1 was inoculated intraperitonealy (i.p.) with tomato extract containing approximately 100 μg of N protein, whereas group 2 was orally immunized with the same amount of N protein-containing tomato protein extract (100 μg). Two groups of five mice each were inoculated i.p. or orally with 100 μg of proteins of non-transformed tomato extract (negative control) or with the rabies vaccine. One booster immunization was given at 30 days post-inoculation (dpi) using 100 μg of an N-protein-enriched chromatographic fraction. Mice were challenge intraperitonealy with rabies virus. The antibody titre of mice immunized i.p. was almost six times higher than that of mice immunized orally. Apparently, protection was dependent on the routes of vaccination as 50% of mice immunized i.p. with 100 μg of N-containing tomato extract were protected whereas all animals immunized orally with plant-derived N died of rabies.

In their third assay, the glycoprotein was expressed using carrots to be used in the immunization of mice (Rojas-Anaya *et al.*, 2009). This plant model was used since the carrot is a vegetable that is widely distributed and easy to produce both *in vivo* as well as *in vitro* and can be consumed raw.

The G gene of the rabies virus arctic fox strain was subcloned between the double enhancer CaMV 35S promoter and CaMV 35S terminator in the vector pUCpSS; this construct was named pUCpSSrabG. For transformation, we decided to use the minimal cassette expression approach (promoter-gene-terminator). We employed carrot seeds for induction of carrot callus. For selection, we employed the gene coding for phosphinothricin acetyl transferase (*bar*), which confers resistance to the herbicide Basta.

Embryogenic calli were transformed by biolistics and herbicide-resistant plants were obtained in liquid medium. The presence of the G gene in leaves was determined by PCR and the protein was detected by Western blot using rabbit polyclonal serum against rabies G protein. In order to evaluate the carrot as vaccine, 24 mice were divided into four groups: G1, fed standard mouse chow (negative control); G2, received an intramuscular dose of inactivated rabies vaccine; G3, mice fed 50 μg of rabies virus G protein in 2 g of raw carrot; G4, mice received 50 μg rabies virus G protein contained in 2 g of raw carrot plus 50 μg of N protein rabies virus (N protein was orally administered) since this molecule has been reported as adjuvant in some rabies vaccines. Mice vaccinated were challenged intracerebrally 60 days post-vaccination.

We were able to regenerate 300 adult plants from 100 calli selected in liquid medium, and 93.3% of the analysed plants showed integration of the transgene with levels of expression varying from 0.2 to 1.4% TSP. In our project, the band corresponding to the recombinant G protein migrated slightly above the native G protein (~70 versus 65 kDa); one likely explanation for this was glycosylation of the protein by the plant. We showed that the ingestion of antigen expressed in carrot results in protective rabies antibodies. These results are consistent with previous studies where the glycoprotein of rabies virus was expressed either in tobacco or in spinach. In this study, we did not observe a 100% protection of the mice; this is possibly because a greater concentration of G protein is necessary. To improve the protective dose, 100 μg (4 g of carrots) could be administered instead of the 50 μg used in this study.

Transgenic maize was recently supplied under controlled conditions at various dosages to sheep via oral administration of a single dose. The results showed that 2 g of the G protein of the rabies virus protected more than 80% of the animals challenged with a lethal vampire origin virus (Loza-Rubio *et al.*, 2012b). This assay used the same conditions for obtaining transformed maize reported previously by Loza-Rubio *et al.* (2008). Similarly, the Basta herbicide was used for selecting transformed plants.

When the plants reached adulthood, kernels expressing the glycoprotein were identified by PCR and Western blot and were subsequently pooled and quantified before immunization. The animals were divided into six groups containing six animals per group, as follows: Group 1, sheep fed 0.5 mg of rabies virus G protein in 20 g of ground maize kernels; Group 2, sheep fed 1.0 mg of rabies virus G protein in 40 g of ground maize kernels; Group 3, sheep fed 1.5 mg of rabies virus G protein in 60 g of ground maize kernels; Group 4, sheep fed 2.0 mg of rabies virus G protein in 80 g of ground maize kernels; Group 5, sheep vaccinated with an inactivated rabies vaccine administered intramuscularly; and Group 6, animals fed 40 g of non-transformed ground maize kernels. Once all groups had been immunized, they were deprived of feed and water for 4 h. The animals were bled to evaluate immune response in serum. Sheep were challenged by the injection at 120 days post-vaccination. The G protein was detected slightly above the native G protein (~70 kDa). The same increase in molecular weight was observed in polyacrylamide gel electrophoresis; the differentially expressed band seems to be heavier than the native G protein (positive control). The expression level obtained from this analysis was an average of 25 μg of G recombinant protein per gram of fresh tissue in the three different lines.

At day 30 post-vaccination, rabies virus antibodies were detected in all vaccinated groups. Animals that received one or two doses of antigen (0.5 and 1.0 mg, respectively) showed a survival rate of 50% (three deaths in six vaccinated animals). In Group 3, which was immunized with 1.5 mg of protein G, only two sheep died of rabies (2/6), with a survival rate of 66% (Fig. 8.4). The lowest mortality (1/6) was found among sheep immunized with the commercial vaccine and those receiving 2.0 mg of protein, which protected 83% of the animals. In this study, a large amount of protein was needed to elicit an immune response because a significant portion of the recombinant protein was likely degraded in the rumen. Although we did not observe any signs of tolerance, this could be because the sheep were fed the edible vaccine only once. These results demonstrated the effectiveness of the oral immunization of sheep with maize that expresses the rabies virus G protein. This is the first study in which an orally administered edible vaccine has shown efficacy in a polygastric species.

This new technology could contribute to global vaccination programmes and have a dramatic impact on public and veterinary health, not only in our country, but also in others with similar problems. Nevertheless, the fact that transformed maize must not be grown on open fields must be taken into account since it is an open-pollination plant. It should be grown in greenhouses with the highest biosecurity. There are also issues that still need to be resolved, such as the antigen dose that each plant produces, since it could produce tolerance (Loza-Rubio and Rojas-Anaya, 2010).

(A)   (B)

Fig. 8.4. Challenge of sheep with a lethal virus of a vampire source, the main rabies transmitter in Latin America. (A) Mature maize plant for immunization and (B) sheep rabid post-challenge (photos courtesy of E. Loza-Rubio).

8.7 Legal Issues

Once the aforementioned goal is achieved, there are other important points that must be covered regarding the design and elaboration of a vaccine derived from plants, such as assuring transgene stability and antigen expression in the following generations, guaranteeing the consistency and reproducibility of the methods used to obtain the vaccine, evaluation and monitoring of the environment to avoid contamination of endemic species, making sure that efficient methods are available for quantification of the antigen to determine correct dosing and bioavailability of the vaccine, and assessment of the handling and transporting techniques with regard to wastes produced from the transformed plants. Because of all this, legislation is necessary, both at global and local levels, on the research and development of biological products using transgenic plants as a platform. In this regard, Hungary became the first Central European country to adopt legislation on the regulation of genetic engineering activity. Its Gene Technology Law entered into force in January 1999 with the concomitant establishment of an advisory body (The Gene Technology Committee). Subsequently, several countries of Central and Eastern Europe adopted regulations with regard to genetically modified organisms.

Before 2000, the differences between the USA and the European Community in their approach to the regulation of biotechnology had arisen because of the different initial cognitive frameworks, a different level of trust in the government and the dissimilar agropolitical situation. The dissimilar cognitive frameworks arise primarily because many Europeans appear to view the environment as a fragile ecosystem that may be easily unbalanced by transgenic plants. In contrast, the dominant view in the USA is of a resilient environment that can easily adapt. In September 2003, the Cartagena Protocol, an international treaty governing the movements of living modified organisms (LMOs) resulting from modern biotechnology from one country to another, was adopted (UN, 2000). One of the principal objectives of the Protocol is to provide information to importing countries to assist their decision-making when accepting imports of LMOs. Now, boasting almost 190 member governments all around the world (known as 'Parties'), the Convention has three goals: the conservation of biodiversity, the sustainable use of the components of biodiversity and the fair and equitable sharing of the benefits arising from the use of genetic resources (Loza-Rubio and Rojas-Anaya, 2010).

During the Global Biotechnology Forum FAO in 2005 (Ruane and Sonino, 2008), countries discussed the kinds of genetically modified organism (GMO) regulatory systems that might be appropriate for developing countries. It is important to consider that GMOs for food and agriculture are a very heterogeneous group, for example the potential environmental risks from GM forest trees and the release of a GM yeast to make bread are different. In addition, within each of these sectors, GMOs may vary considerably, requiring different kinds of regulations, for example:

- Some species are not grown for food (e.g. cotton), so food safety regulations are not strictly an issue. However, it should be kept in mind that some material, e.g. pollen/honey derived from GMOs, may still enter the food chain.
- The same species may be modified for very different traits, e.g. an agricultural crop or animal may be modified to produce human pharmaceuticals such as tomatoes producing vaccines against virus or animals producing hormones. 'Pharmed' products under development include vaccines, antibodies and industrial proteins and in the crop sector, involve banana, maize, potato and tomato plants. Special regulations covering potential gene flow to their conventional counterparts may be necessary.
- Regulations may vary depending on whether the GM species is produced for export or domestic use. For this reason, GMO commercialization is subject to a strict marketability requirement. Otherwise, GMO varieties are not approved for commercialization. When exports are not a significant factor (e.g. in the case of cotton), commercial release can be approved irrespective of the regulatory status elsewhere, since there are no 'sensitive' markets for 'the product'.

On the other hand, there are 12 countries that are noteworthy due to their high levels of biodiversity: Brazil, Indonesia, Colombia, Australia, Mexico, Madagascar, Peru, China, Philippines, India, Ecuador and Venezuela, known as mega-diverse, therefore it is very important to preserve their indigenous germplasm (Tovar, 2008).

The identity of Mexico in terms of its biodiversity is important, therefore its global role must be recognized. Mexico is a country with domestic and wild plant species of which it is their centre of origin, so their protection has been requested. In this context, Mexico is one of the countries that are signatory to the Cartagena Protocol that came into effect in 2003 (UN, 2000). The majority of the problems related to gene flow corresponding to GMOs and the issues regarding the responsibility/compensation were examined in the legislation framework of the Cartagena Protocol on Biosafety to the Convention on Biological Diversity. Article 1 mentions:

> In accordance with the precautionary approach contained in Principle 15 of the Rio Declaration on Environment and Development, the objective of this Protocol is to contribute to ensuring an adequate level of protection in the field of the safe transfer, handling and use of living modified organisms resulting from modern biotechnology that may have adverse effects on the conservation and sustainable use of biological diversity, taking also into account risks to human health, and specifically focusing on transboundary movements.

Before 2005, the Ecological Equilibrium and Environmental Protection General Act (Tovar, 2008) was the national judicial instrument that provided the basis for regulation regarding GMOs. It has the objective of 'regulating the activities of confined use, experimental release, pilot program release, commercial release, marketing, import and export of genetically modified organisms' in order to prevent, avoid or reduce the possible risks that these activities could cause to human health, as well as to the health of animals, plants and water animals, the environment and the biological diversity of the country. It establishes as competent authorities for issuing permits and sanctions the Ministry of the Environment and Natural Resources, The Ministry of Agriculture, Livestock, Rural Development, Fisheries and Food and the Ministry of Health. It also establishes the basis for the operation of the Inter-ministry Commission on Biosafety for Genetically Modified Organisms (CIBIOGEM) through which the various aforementioned ministries must collaborate regarding the biosafety of GMO.

The Law established that a permit will be required for carrying out the following activities (LBOGM, 2005):

1. Experimental release into the environment, including imports for this activity, of one or more GMOs.
2. Release into the environment in a pilot programme, including imports for this activity.
3. The commercial release into the environment, including imports for this activity, of GMOs.

In order to establish a risk assessment of the aforementioned points each case is to be analysed individually through scientific and technical studies carried out by the interested parties, evaluating the possible risks of the experimental release into the environment and to the biological diversity, as well as to the health of animals, plants and fisheries. The studies must include possible risks to human health. Up to the development of this document no reference has been made regarding transformed plants that express an antigen (plant-derived antigens).

The latest amendment to the Regulations of the Genetically Modified Organisms Biosafety Act was published in 2009 (Reglamento de la Ley de Bioseguridad de Organismos Geneticamente Modificados, 2009). These regulations establish, among other things:

1. The characteristics that must be contained within the request for permission to carry out activities using GMOs.
2. The requirements for permits for release into the environment.
3. Considerations on the import and export of GMOs that are destined for their release into the environment.
4. Characteristics of the Internal Biosafety Commissions of public and private institutions.
5. Determination of the centres of origin and genetic diversity.
6. Establishment of The National Biosafety Information System.
7. Determination of the list of GMOs that are to be issued by the competent Ministries.

One of the most important publications of the regulations is the 'Special protection regime for corn'. In the context of plant-derived vaccines, it is important to mention that the regime establishes: 'The experimentation, or release into the environment of genetically modified corn that has characteristics that prevent or limit its use or consumption by humans or animals shall not be allowed, as well as their use in the processing of food for human consumption'. The aforementioned blocks the experimentation with maize for the development of antigens and other biopharmaceuticals.

On the other hand, Brazil, another mega-diverse country with great advances in biotechnology development, promulgated Decree 6.041 of the Policy on the Development of Biotechnology (National Biotechnology Committee, 2007), the objective of which is:

> To promote and carry out actions in order to establish the adequate environment for developing biotechnology products and innovative processes, promote the greatest efficiency of the national productive structure, the innovative capacity of Brazilian companies, the adsorption of technologies, the generation of business and the expansion of exports.

In comparison with the Mexican regulation it also established the competences of the Ministries regarding activities with GMOs, the integration of Biosafety Committees, etc. It is noteworthy that in said document is established as a strategic objective to stimulate the production of recombinant proteins using plants, animals and microorganisms as bioreactors, and the plants resistant to biotic and abiotic stress. The aforementioned places emphasis on the coexistence of transgenic and conventional varieties promoting the development of mechanisms and technologies for preserving the genetic identity of cultivars, as well as the development of geographical information systems for monitoring and zoning of the activities related to distance biotechnology safety.

Furthermore, technological development in Peru into biotechnology is governed by the Modern Biotechnology Development of Peru General Act published in July 2006, the objective of which is:

> to promote modern biotechnology through scientific research and technological development and innovation; increase competitiveness, economic development and the wellbeing of the population, in harmony with human health and with the preservation of the environment.

This law, similar to the Brazilian law, has some similarity with Mexican regulations such as the protection of native species, the constitution of biosafety committees and inter-ministry commissions on biotechnology, whose mission is to harmonize policies in the area of biotechnology. It also has articles regarding intellectual property of biotechnology-derived products. The Peruvian law regarding modern biotechnology safety (Law 27104) as well as its regulations, in force since 2003, establishes that an adequate level of protection must be guaranteed to human health, the environment, biological diversity and their sustainable use in the area of generation, research, production, transfer, handling, transport, storage, conservation, exchange, marketing, confined use and release into the environment of GMOs and their by-products. They establish a risk assessment of each individual case, as well as of the labelling of GMO by-products. The analysis of risks derived from gene flow in the development of genetically modified organisms is promoted. The risk assessment shall focus on the characteristics of the GMOs and their by-products.

The Argentine Regulatory Framework dates back to 1991, when Argentina started its incursion into biotechnology (SAGPyA, 2004). At this time the objective was to guarantee that GMOs that are released into the environment or for human consumption are safe for the agroecosystem, safe for human and animal consumption, and are convenient for the country from a commercial point of view. In 2002, the 'National framework on GMO biosafety in Argentina' was published. This country does not have a specific law regarding GMO biosafety, but it does have a regulatory framework, and is a signatory to the Cartagena Protocol. Guidelines are established for handling and assessment of GMOs in confinement and release conditions, the considerations for forming the evaluation and biosafety committees, the approval for their use as food stuff, the ruling on the safety of GMO derived food, as well as contemplates the case-by-case risk analysis of the non-GMO counterpart (SAGPyA/UNEP–GEF, 2004).

Recent developments in genetic modification and the use of LMOs in agriculture have ignited a debate over the potential effects of these organisms on biological diversity. The regime does allow states to enact national protective measures to preserve human and animal health as well as natural resources, based on scientific evidence. However, it is necessary to ensure that this is not only on paper, but is carried out in order to avoid ecological imbalances that could affect all species, including humans.

References

Arellano-Sota, C. (1988) Vampire-bat transmitted rabies in cattle. *Review of Infectious Diseases* 10, 707–709.

Ashraf, S., Singh, P.K., Yadav, D.K., Shahnawaz, M., Mishra, S. and Sawant, S.V. (2005) High level expression of surface glycoprotein of rabies virus in tobacco leaves and its immunoprotective activity in mice. *Journal of Biotechnology* 119, 1–14.

Brochier, B., Aubert, M.F., Pastoret, P.P., Masson, E., Schon, J., Lombard, M., Chappuis, G., Languet, B. and Desmettre, P. (1996) Field use of a vaccinia-rabies recombinant vaccine for the control of sylvatic rabies in Europe and North America. *Revue Scientifique et Technique* 15, 947–970.

Cadoz, M., Strady, A., Meigner, B., Taylor, J., Tartaglia, J., Paoletti, E. and Plotkin, S. (1992) Immunisations with canarypox virus expressing rabies glycoprotein. *The Lancet* 339, 1429–1432.

Delpietro, H.A., Lord, R.D., Russo, R.G. and Gury-Dhomen, F. (2009) Observations of sylvatic rabies in Northern Argentina during outbreaks of paralytic cattle rabies transmitted by vampire bats (*Desmodus rotundus*). *Journal of Wildlife Disease* 45, 1169–1173.

Dietzgen, R.G., Calisher, C.H., Kurath, G., Kusmin, I.V., Rodríguez, L.L. and Stone, D.M. (2011) Family Rhabdoviridae. In: King, A.M., Adams, M.J., Carstens, E.B. and Lefkowitz, E.J. (eds) *Virus taxonomy: classification and nomenclature of viruses.* Ninth Report of the International Committee on Taxonomy of Viruses. Elsevier, San Diego, California.

Food and Agriculture Organization of the United Nations (2011) *Investigating the role of bats in emerging zoonoses: Balancing ecology, conservation and public health interests* (eds Newman, S.H., Field, H.E., de Jong, C.E. and Epstein, J.H.). FAO Animal Production and Health, Rome.

Freuling, C.M., Beer, M., Conraths, F.J., Finke, S., Hoffmann, K.B., Kiemt, J., Mettenleiter, T.C., Mühlbach, E., Teifke, J., Wohlsein, P. and Müller, T. (2011) Novel Lyssavirus in Nattere's Bat in Germany. *Emerging Infectious Diseases* 17, 1519–1522.

Fry, T.L., Vandalen, K.K., Shriner, S.A., Moore, S.M., Hanlon, C.A. and Vercauteren, K.C. (2013) Humoral immune response to oral rabies vaccination in raccoon kits: Problems and implications. *Vaccine* doi: 10.1016/j.vaccine.2013.04.016.

Greenhall, A.M. (1993) *Bats and Rabies.* Edition Fondation Marcel Mérieux, Lyon, France.

Hanlon, C.A., Niezgoda, M., Morrill, P. and Rupprecht, C.E. (2002) Oral efficacy of an attenuated rabies virus vaccine in skunks and raccoons. *Journal of Wildlife Disease* 38, 420–427.

Henderson, H., Jackson, F., Bean, K., Panasuk, B., Niezgoda, M., Slate, D., Li, J., Dietzschold, B., Mattis, J. and Rupprecht, C.E. (2009) Oral immunization of raccoons and skunks with a canine adenovirus recombinant rabies vaccine. *Vaccine* 27, 194–197.

Hooper, D.C., Pierrard, L., Modelska, A., Otvos, L., Fu, Z., Koprowski, H. and Dietzschold, B. (1994) Rabies nucleocapsid as an oral immunogen immunological enhancer. *Proceedings of the National Academy of Sciences USA* 91, 10908–10911.

Jackson, C.A. and Wunner, H.W. (2007) Rabies virus. In: Jackson, C.A. and Wunner, H.W. (eds) *Rabies,* 2nd edn. Elsevier, London, pp. 45–47.

Johnson, N., Cunningham, A.F. and Fooks, A.R. (2010) The immune response to rabies virus infection and vaccination. *Vaccine* 28, 3896–3901.

Kieny, M.P., Lathe, R., Drillien, R., Spenhner, D., Skory, S., Schmitt, D., Wiktor, T., Koprowski, H. and Lecocq, J.P. (1984) Expression of rabies virus glycoprotein from a recombinant vaccinia virus. *Nature* 312, 163–166.

Kuzmin, I.V., Mayer, A.E., Niezgoda, M., Markotter, W., Agwanda, B., Breiman, R.F. and Rupprecht, C.E. (2010) Shimoni bat virus, a new representative of the Lyssavirus genus. *Virus Research* 149, 197–210.

Lafon, M. (2005) Rabies virus receptors. *Journal of Neurovirology* 11, 82–87.

Lafon, M. (2011) Evasive strategies in rabies virus infection. *Advances in Virus Research* 79, 33–53.

Lamphear, B.J., Streatfield, S.J., Jilka, J.M., Brooks, C.A., Barker, D.K., Turner, D.D., Delaney, D.E., Garcia, M., Wiggins, B., Woodard, S.L., Hood, E.E., Tizard, I.R., Lawhorn, B. and Howard, J.A. (2002) Delivery of subunit vaccines in maize seed. *Journal of Controlled Release* 85, 169–180.

LBOGM (2005) Ley de Bioseguridad de Organismos Geneticamente Modificados. Diario Oficial de la Federación. México.

Loza-Rubio, E. and Rojas-Anaya, E. (2010) Vaccine production in plant systems – an aid to the control of viral diseases in domestic animals: A review. *Acta Veterinaria Hungarica* 58, 511–522.

Loza-Rubio, E., Vargas, G., Hernández, E., Batalla, D. and Aguilar-Setien, A. (1996) Investigation of rabies virus strains in Mexico with a panel of monoclonal antibodies used to classify Lyssaviruses. *Bulletin of the Pan American Health Organization* 30, 31.

Loza-Rubio, E., Pedroza-Requénes, R., Montano-Hirose, J.A. and Aguilar Setién, A. (1998) Caracterización con anticuerpos monoclonales de virus de la rabia aislados de fauna doméstica y silvestre de México. *Veterinaria México* 29, 345–349.

Loza-Rubio, E., Rojas, A.E., Banda, R.V.M., Nadin-Davis, S.A. and Cortez, G.B. (2005) Detection of multiple strains of rabies virus RNA using primers designed to target Mexican vampire bat variants. *Epidemiology and Infection* 133, 927–934.

Loza-Rubio, E., Rojas, A.E., Gómez, N.L., Olivera, F.M.T.J. and Gómez-Lim, M. (2008) Development of an edible rabies vaccine in maize using the Vnukovo strain. *Developments in Biologicals* 131, 477–482.

Loza-Rubio, E., Molina, G.I. and Montaño, H.J.A. (2009) Nucleocapsid of rabies virus improve immune response of an inactivated avian influenza vaccine. *Veterinary Research Communications* 33, 589–595.

Loza-Rubio, E., Nadin-Davis, S.A. and Morales, S.E. (2012a) Molecular and biological properties of rabies viruses circulating in Mexican Skunks: focus on P gene variation. *Revista Mexicana de Ciencias Pecuarias* 3, 155–170.

Loza-Rubio, E., Rojas, A.E., Gómez, N.L., Flores, O.M.T. and Gómez-Lim, M.A. (2012b) Protein G from rabies virus expressed in maize protects mice against a challenge with an heterologous rabies virus. *Vaccine* 30, 5551–5556.

Mainguy, J., Fehlner-Gardiner, C., Slate, D. and Rudd, R.J. (2013) Oral rabies vaccination in raccoons: comparison of ONRAB® and RABORAL V-RG® vaccine-bait field performance in Québec, Canada and Vermont, USA. *Journal of Wildlife Diseases* 49, 190–193.

Marston, D.A., Horton, D.L., Ngeleja, C., Hampson, K., McElhinney, L.M., Banyard, A.C., Haydon, D., Cleaveland, S., Rupprecht, C.E., Bigambo, M., Fooks, A.R. and Lembo, T. (2012) Ikoma lyssavirus, highly divergent novel lyssavirus in an African civet. *Emerging Infectious Disease* 18, 664–667.

McGarvey, P.B., Hammond, J., Dienelt, M.M., Hooper, D.C., Fu, Z.F., Dietzschold, B., Koprowski, H. and Michaels, F.H. (1995) Expression of the rabies virus glycoprotein in transgenic tomatoes. *Biotechnology* 13, 1484–1488.

Modelska, A., Dietzschold, B., Fleysh, N., Fu, Z.F., Steplewski, K., Hooper, C. and Koprowski, H. (1998) Immunization against rabies with plant-derived antigen. *Proceedings of the National Academy of Sciences USA* 95, 2481–2485.

Modern Biotechnology Development of Peru General Act (2006) Ley de Desarrollo de la Biotecnología Moderna en el Perú. Presidencia de la Republica.

Morales, M.M., Rico, R.G., Gómez, O.J. and Aguilar, S.A. (2006) Importancia inmunológica de la proteína N en la infección por virus de la rabia. *Veterinaria México* 37(3), 352–364.

Nadin-Davis, S.A., Abdel-Malik, M., Armstrong, J. and Wandeler, A.I. (2002) Lyssavirus P gene characterization provides insights into the phylogeny of the genus and identifies structural similarities and diversity within the encoded phosphoprotein. *Virology* 298, 286–305.

National Biotechnology Committee (2007) Biotechnology Development Policy. National Biotechnology Committee, Brazil.

Perea-Arango, I., Loza-Rubio, E., Rojas-Anaya, E., Olivera, F.M.T.J., De la Vara, G.L. and Gómez, L.M. (2008) Expression of rabies virus nucleoprotein in plants at high levels and evaluation of immune response in mice. *Plant Cell Reports* 27(4), 677–685.

Reglamento de la Ley de Bioseguridad de Organismos Geneticamente Modificados (2009) Dioario Oficial de la Federación.

Rojas-Anaya, E., Loza-Rubio, E., Flores, O.M.T. and Gómez-Lim, M.A. (2009) Expression of rabies virus G protein in carrots (*Daucus carota*). *Transgenic Research* 18, 911–919.

Romero, A.M.L., Aguilar, S.À. and Sánchez, H.C. (2006) *Murciélagos benéficos y vampiros: características, importancia, rabia, control y conservación.* AGT Editor, S.A. México DF, pp.75–78.

Ross, B.A., Favi, M.C. and Vásquez, V. (2008) Glicoproteína del virus rábico: Estructura, inmunogenicidad y rol en la patogenia. *Revista Chilena de Infectología* 25, S14–18.

Ruane, J. and Sonino, A. (2008) Results from the FAO Biotechnology Forum. FAO, Rome.

Rybicki, E.P. (2009) Plant-produced vaccines: promise and reality. *Drug Discovery Today* 14, 16–24.

SAGPyA (2004) Plan Estratégico 2005-2012 para el Desarrollo de la Biotecnología Agropecuaria. Secretaria de Agricultura, Ganadería, Pesca y Alimentos de la República Argentina.

SAGPyA/UNEP–GEF (2004) Revisión del Marco Nacional sobre Bioseguridad en Argentina. Secretaria de Agricultura, Ganadería, Pesca y Alimentos de la República Argentina.

Sakamoto, S., Ide, T., Tokiyoshi, S., Nakao, J., Hamada, F., Yamamoto, M., Grosby, J.A., Ni, Y. and Kawai, A. (1999) Studies on the structures and antigenic properties of rabies virus glycoprotein analogues produced in yeast cells. *Vaccine* 17, 205–218.

Shors, T. (2009) *Virus: estudio molecular con orientación clínica.* Editorial Médica Panamericana, S.A., p. 367.

Streatfield, S.J. (2006) Mucosal immunization using recombinant plant-based oral vaccines. *Methods* 38, 150–157.

Tacket, C. (2009) Plant-based oral vaccines: results of human trials. *Current Topics in Microbiology and Immunology* 332, 103–117.

Tovar, M.P. (2008) Ley General del Equilibrio ecologico y la Protección al Ambiente. In: *Bioseguridad en la aplicación de la Biotecnología y el uso de los organismos geneticamente modificados.* CIBIOGEM, pp. 287.

United Nations (2000) *Cartagena Protocol on Biosafety to the Convention on Biological Diversity.* United Nations, Montreal.

Ventini, D.C., Astray, R.M., Jemos, M.A., Jorge, S.A., Riquelme, C.C., Suazo, C.A., Tonso, A. and Pereira, C.A. (2010) Recombinant rabies virus glycoprotein synthesis in bioreactor by transfected *Drosophila melanogaster* S2 cells carrying a constitutive or an inducible promoter. *Journal of Biotechnology* 146, 169–172.

Woldehiwet, Z. (2002) Rabies: recent developments. *Research in Veterinary Science* 73, 17–25.

World Health Organization (2008) WHO expert consultation on rabies. Factsheet no. 99. WHO, Geneva.

Xiang, Z.Q., Yang, Y., Wilson, J.M. and Ertl, H.C.J. (1996) A replication-defective human adenovirus recombinant serves as highly efficacious vaccine carrier. *Virology* 219, 220–227.

Yusibov, V., Modelska, A., Steplewski, K., Agadjanyan, M., Weiner, D., Hooper, C. and Koprowski, H. (1997) Antigens produced in plants by infection with chimeric plant viruses immunize against rabies virus and HIV-1. *Proceedings of the National Academy of Sciences USA* 94, 5784–5788.

Yusibov, V., Hooper, D.C., Spitsin, S.V., Fleysh, N., Kean, R.B., Mikheeva, T., Deka, D., Karasev, A., Cox, S., Randall, J. and Koprowski, H. (2002) Expression in plants and immunogenicity of plant virus-based experimental rabies vaccine. *Vaccine* 20, 3155–3164.

9 Case Study 2: Plant-made HIV Vaccines and Neutralizing Antibodies

Carla Marusic* and Marcello Donini

ENEA, Laboratorio Biotecnologie, UTBIORAD, Rome, Italy

9.1 Introduction

The latest UNAIDS 2012 global report (http://www.unaids.org) showed that a total of 34 million people were living with human immunodeficiency virus (HIV). Although data show a 50% reduction in the rate of new HIV infections (HIV incidence) between 2001 and 2011 in 25 low- and middle-income countries, a safe, effective and durable HIV vaccine remains a highest priority. Along with the development of efficacious preventive vaccines, therapeutic treatments designed to control HIV infections in people who are already HIV positive are needed and crucial. In this context, antibodies with broad neutralizing activity have been proposed for both prophylaxis and therapy (Abela *et al.*, 2010).

As the majority of the new HIV infections occur in sub-Saharan Africa, novel low-cost production systems that allow the establishment of on-site manufacturing facilities would have a tremendous impact in vaccine development. With the first FDA approved plant-derived pharmaceutical on the market for the treatment of Gaucher's disease (Maxmen, 2012), plants could represent the ideal, safe and economic alternative to traditional production systems for the large-scale production of HIV vaccine components and biopharmaceuticals. In particular, a large amount of data published in the last 15 years showed that plant-based strategies are a valid approach for the production of anti-HIV-1 recombinant subunit vaccines. Moreover, plant-produced viral components demonstrated to be effective in a prime-boost strategy in which animals were primed with a DNA vaccine and the plant-derived antigen was used as a boost (Scotti *et al.*, 2010; Rosales-Mendoza *et al.*, 2012).

HIV antigens that are considered promising candidates as vaccine components belong to early-stage, structural and envelope protein groups. Several viral components such as epitopes, single proteins, multi-antigens or virus-like particles (VLPs) have recently been successfully expressed in plants by stable nuclear or plastid transformation or by transient expression systems based on plant virus vectors or *Agrobacterium*-mediated infection

* Corresponding author, E-mail: carla.marusic@enea.it

(Table 9.1). Moreover, the production and characterization of plant-derived anti-HIV neutralizing antibodies is well supported (Table 9.2) and in July 2011 a neutralizing antibody (P2G12), obtained in tobacco plants, entered clinical trial (http://www.pharma-planta.net). This chapter provides an up-to-date overview of plant-based anti-HIV strategies for the production of both vaccine components and neutralizing antibodies.

9.2 Plant-made HIV Vaccine

9.2.1 Functional proteins

Nef

The accessory protein Nef is considered a promising candidate for the formulation of HIV multicomponent vaccines as it is expressed in the early stages of the viral life cycle and is involved in both high viral load and disease progression. Moreover, mutations in the *nef* genes are associated with long-term non-progression patients (Cruz *et al.*, 2013).

Nef is a small cytosolic protein that can be found in the cells as two main isoforms translated by two in-frame translational start codons, p27 (27 kDa), which represents the full-length viral protein, and p25 (25 kDa), a truncated form translated from the second start codon and lacking the first 18 amino acids. p27 is post-translationally modified by the addition of a myristoyl group to the N-terminus, which anchors Nef to the cytosolic side of cellular membranes (Geyer *et al.*, 2001). Both isoforms (p25, p27) have been expressed in different biological systems such as *Escherichia coli* (Azad *et al.*, 1994), yeast (Macreadie *et al.*, 1998; Sirén *et al.*, 2006), insect cells (Kohleisen *et al.*, 1996) and mammalian cells (Cooke *et al.*, 1997). Recently, different biotechnological approaches were developed to obtain high accumulation of Nef proteins in plant cells (Marusic *et al.*, 2009).

Our research group published in 2007 the first study regarding the expression of full-length myristoylated (p27), non-myristoylated (p27 mut) and truncated forms (p25) of Nef either in the cytosol or cell secretory pathway. Transgenic tobacco plants expressing p25 and p27 mut were generated showing an average accumulation of 0.5% of total soluble proteins (TSP) (Marusic *et al.*, 2007). Most notably, the use of a recombinant non-myristoylated Nef is considered advantageous in the production of a multicomponent vaccine as the deletion or mutagenesis of the myristoylation site abrogates Nef mediated down-regulation of both major histo-compatibility complex (MHC) class I and CD4 cell-surface molecules (Peng and Robert-Guroff, 2001; Peng *et al.*, 2006), which typically prevent cytotoxic T lymphocyte (CTL)-mediated lysis of HIV-1-primary infected cells (Collins *et al.*, 1998). Two years later, we devised a strategy to improve Nef accumulation in plants (1.5% TSP) using an *Agrobacterium*-mediated transient expression system. This improvement permitted the development of an optimized purification protocol of plant-produced Nef (Lombardi *et al.*, 2009).

Table 9.1. Plant-expressed HIV-1 antigens.

Protein/Peptide	Host	Expression system	References
Tat	Spinach	TMV plant viral vector	Karasev *et al.*, 2005
	Potato tuber	Nuclear transformation	Kim *et al.*, 2004a
	Tomato	Nuclear transformation	Ramírez *et al.*, 2007
	Tobacco	Nuclear transformation	Webster *et al.*, 2005
	Tomato	Nuclear transformation	Cueno *et al.*, 2010
Nef	Tobacco	Nuclear transformation	Marusic *et al.*, 2007; Barbante *et al.*, 2008; de Virgilio *et al.*, 2008
	Tobacco	Plastid transformation	McCabe *et al.*, 2008; Zhou *et al.*, 2008
	Tomato	Plastid transformation	Zhou *et al.*, 2008; Gonzalez-Rabade *et al.*, 2011
p24	Tobaccc	Nuclear transformation	Obregon *et al.*, 2006
	Nicotiana benthamiana	TBSV plant viral vector	Zhang *et al.*, 2000
	Tobacco	Nuclear transformation	Zhang *et al.*, 2002
	N. benthamiana	TMV plant viral vector	Meyers *et al.*, 2008
	N. benthamiana	Agro-infiltration	Meyers *et al.*, 2008
	Tobacco	Nuclear transformation	Meyers *et al.*, 2008
	Tobacco	Plastid transformation	Meyers *et al.*, 2008
	Tobacco	Plastid transformation	McCabe *et al.*, 2008; Zhou *et al.*, 2008
	Tomato	Plastid transformation	Zhou *et al.*, 2008; Gonzalez-Rabade *et al.*, 2011
	N. benthamiana	TMV plant viral vector	Pérez-Filgueira *et al.*, 2004
	Arabidopsis thaliana	Nuclear transformation	Lindh *et al.*, 2008
p17/p24	*N. benthamiana*	Agro-infiltration	Meyers *et al.*, 2008
	Tobacco	Nuclear transformation	Meyers *et al.*, 2008
Pr55Gag	*N. benthamiana*	TMV plant viral vector	Meyers *et al.*, 2008
	N. benthamiana	Agro-infiltration	Meyers *et al.*, 2008
	Tobacco	Nuclear transformation	Meyers *et al.*, 2008
	Tobacco	Plastid transformation	Scotti *et al.*, 2009
	N. benthamiana	TMV plant viral vector	Kessans *et al.*, 2013
	N. benthamiana	Nuclear transformation	Kessans *et al.*, 2013
Env/gp120			
V3 loop	*N. benthamiana*	TMV plant viral vector	Yusibov *et al.*, 1997
	Tobacco	TMV plant viral vector	Sugiyama *et al.*, 1995; Beachy *et al.*, 1996
	Potato	Nuclear transformation	Kim *et al.*, 2004b
	N. benthamiana	TBSV plant viral vector	Joelson *et al.*, 1997
C4V3	Tobacco	Plastid transformation	Rubio-Infante *et al.*, 2012
Env/gp41			
2F5 epitope (662–667)	*N. benthamiana*	PVX plant viral vector	Marusic *et al.*, 2001
P1 peptide (649–684)	*N. benthamiana*	Agro-infiltration	Matoba *et al.*, 2004, 2009
Peptide (731–752)	Cowpea	CPMV plant viral vector	Porta *et al.*, 1994; McLain *et al.*, 1995, 1996; Durrani *et al.*, 1998; McInerney *et al.*, 1999

TMV, tobacco mosaic virus; TBSV, tomato bushy stunt virus; PVX, potato virus X; CPMV, cowpea mosaic virus.

Table 9.2. Plant produced anti-HIV mAbs.

Expression system	Host	Highest yield	References
2G12			
Transgenic-nuclear	Tobacco	Not reported	http://www.pharma-planta.net
Transgenic-nuclear	Maize (seeds)	60 µg g^{-1} DW	Rademacher et al., 2008
Transgenic-nuclear	Maize (seeds)	100 µg g^{-1} DW	Ramessar et al., 2008
Transient (agro-infiltration)	Nicotiana benthamiana	50 µg g^{-1} FW	Strasser et al., 2008
Transient (CPMV-based vector)	N. benthamiana	100 µg g^{-1} FW	Sainsbury et al., 2010
Transient (agro-infiltration)	N. benthamiana	100 µg g^{-1} FW	Rosenberg et al., 2013
Transgenic-nuclear	Tobacco cells BY-2	8 µg ml^{-1}	Holland et al., 2010
Transgenic-nuclear (2G12 fused with elastin-like peptides)	Tobacco	1% TSP	Floss et al., 2009
2F5			
Transgenic-nuclear	Tobacco cells BY-2	6.4 µg g^{-1} WCW	Sack et al., 2007
Transgenic-nuclear	Maize (seeds)	0.6 µg ml^{-1} seed extract	Sabalza et al., 2012
Transgenic-nuclear (2F5 fused with elastin-like peptides)	Tobacco	0.6% TSP	Floss et al., 2008
4E10			
Transgenic-nuclear	Tobacco (rhyzosecretion in the culture medium)	10.43 mg g^{-1} root DW	Drake et al., 2009
Transgenic-nuclear	Tobacco	Not reported	Platis et al., 2009
Transient (agro-infiltration)	N. benthamiana	Not reported	Rosenberg et al., 2013
Transient (TMV-based vector)	N. benthamiana	Not reported	Strasser et al., 2009
b12			
Transgenic-nuclear	Tobacco	Not reported	Sexton et al., 2009
VRC01			
Transient (agro-infiltration)	N. benthamiana	600 µg g^{-1} FW	Rosenberg et al., 2013
Transient (TMV-based vector)	N. benthamiana	150 µg g^{-1} FW	Hamorsky et al., 2013

WCW, wet cell weight; FW, fresh weight; DW, dry weight; CPMV, cowpea mosaic virus; TSP, total soluble protein; TMV, tobacco mosaic virus.

Other alternative strategies were devised to improve p27 accumulation in transgenic plants, one of which involved the fusion of the C-terminal portion of the mammalian ER (endoplasmic reticulum) isoform of cytochrome b5 (ER *b5*) to Nef, in order to anchor the molecule to the ER cytosolic face. Resulting expression levels were in the range of 0.7% of TSP with very little variability within different tobacco lines (Barbante *et al.*, 2008). Another approach, described by de Virgilio and colleagues, directed p27 accumulation into the plant secretory pathway in order to promote the formation of protein bodies (PB) containing Nef (de Virgilio *et al.*, 2008).

Interesting results were obtained by plastid transformation of tobacco and tomato plants producing Nef as a fusion protein with HIV p24 antigen (Zhou *et al.*, 2008). Analysis of p24-Nef and Nef-p24 fusion proteins showed that both could be expressed in plants at high levels. However, the best results in terms of protein expression levels (close to 40% of the leaf's total protein) were obtained with the p24-Nef fusion protein. Three years later, the immunogenicity of chloroplast-derived p24-Nef fusion protein was reported. In this work, mice were used to evaluate the oral immunogenicity of chloroplast-derived p24-Nef compared to *E. coli*-produced p24 or Nef, administered by oral gavage. The results showed that after three doses of plant-derived p24-Nef in combination with the cholera toxin B subunit (CTB) used as an adjuvant, no immune response to either p24 or Nef was detectable in mouse sera. However, a boost with chloroplast-derived p24-Nef fusion protein after subcutaneous injection with *E. coli*-produced p24 or Nef, was able to elicit a strong specific IgG response to both proteins (Gonzalez-Rabade *et al.*, 2011). Overall data indicate that plant-produced Nef could be effective in a prime-boost vaccination strategy.

Tat

HIV-1 Tat regulatory protein is a small nuclear molecule (86–101 amino acid residues) involved in enhancing the efficiency and initiation of viral transcription. Despite the high viral mutation rate of HIV-1, Tat is a well-conserved protein, particularly within its N-terminal residues (Debaisieux *et al.*, 2012). A substantial amount of evidence suggests that extracellular Tat acts as a viral toxin that affects the biological activity of different cell types and has a key role in acquired immune-deficiency syndrome development (Cafaro *et al.*, 1999), suggesting that Tat could be a good candidate for the development of an effective multicomponent anti-HIV-1 vaccine.

Like Nef, Tat has also been expressed in plants as full-length or as a fusion protein in the attempt to improve its accumulation or immunogenicity. The first study concerning the expression of Tat fused to CTB in transgenic potato was published in 2004. In this work, the sequence encoding the simian-human immunodeficiency virus (SHIV) Tat was fused to the 3' end of the cDNA encoding the CTB subunit, and the derived DNA construct was used to transform *Solanum tuberosum* cells. Although CTB-Tat fusion protein was expressed at low levels (0.005–0.007% TSP), its biological activity was demonstrated (Kim *et al.*, 2004a). However, Kim and colleagues did not perform animal oral immunization studies with transgenic tubers. A different plant expression strategy was used to express Tat in tobacco plants. In this case, two *tat* genes encoding a truncated and the full-length protein (72 and 101 amino acids, respectively) were cloned under the control of the tobacco Pr1 signal peptide and the KDEL ER-retention signal sequence. Tobacco-expressed Tat was used to immunize mice intra-peritoneally with or without adjuvants, but no specific antibodies to Tat were detected (Webster *et al.*, 2005). One year later, the oral delivery of plant-expressed Tat was approached by transient transformation of spinach leaves. Mice were fed with 1 g of inoculated leaves (at days 0, 7 and 21) and the immunological properties of oral delivered Tat were assessed by ELISA. A peak in antibody titres was shown only in mice that received DNA vaccination after oral priming (Karasev *et al.*, 2005). In 2007 the expression of HIV-1 Tat in tomato fruit and its immunogenicity was reported. Different groups of female mice were immunized (at days 0, 14 and 28) orally, intraperitoneally or intramuscularly with tomato fruit extract and boosted after 15 weeks. All immunization routes elicited a strong specific anti-Tat immune response. Moreover, antibody isotyping showed the presence of IgA in oral immunized animals that plays a critical role in mucosal immunity. The functional activity of antibodies in neutralizing the extracellular Tat was assessed by an *in vitro* neutralization assay (Pena-Ramírez *et al.*, 2007). Tat was also expressed in tomato plants fused to GUS protein (β-glucuronidase). In this work, plant extracts were used to inject mice intradermally and both Tat-specific antibody response and cytotoxic T lymphocyte (CTL) activity was shown (Cueno *et al.*, 2010). Although still preliminary, these data indicate that oral delivery of Tat could represent a possible strategy in future vaccine development.

9.2.2 Structural proteins

Pr55gag, gag, p24 and p17–p24

The ideal anti-HIV vaccine formulation should be able to induce a protective immune response by the induction of broadly virus-specific cytotoxic T lymphocytes. Among HIV-1 components, Gag precursor Pr55gag, as well as the proteins derived from its sequential processing (p17 and p24), contain the highest density of CTL epitopes (Novitsky *et al.*, 2002; Addo *et al.*, 2003). Moreover, Pr55gag expressed in different cell systems is able to assemble into non-infectious VLPs which are morphologically comparable to immature virions and capable of eliciting strong cellular and humoral immune responses (Doan *et al.*, 2005).

Several studies showed the production of Pr55gag, p24 and p17 structural proteins in plant tissues. In the first study published in 2000 (Zhang *et al.*, 2000), the *p24* gene was engineered to express the viral protein as an in-frame fusion protein with the N-terminus of tomato bushy stunt virus (TBSV) coat protein (CP) in *Nicotiana benthamiana*. The plant-produced antigen was recognized by p24-specific antibodies, showing the preservation of relevant antigenic determinants. The yield of protein in infected leaves estimated by Western blot was at 5% of TSP. Two years later, the same authors described the expression of p24 in transgenic tobacco plants, but no data concerning immunogenic properties were reported (Zhang *et al.*,

2002). P24-histidine (His)-tagged protein was transiently expressed in tobacco and purified by immobilized metal affinity chromatography (IMAC) (Pérez-Filgueira *et al.*, 2004). The immunological potential of plant-produced HIV-1 p24 was demonstrated by Obregon and colleagues (Obregon *et al.*, 2006). In this work, *p24* and *IgA* genes were engineered to express an antigen-antibody fusion protein in the ER of transgenic tobacco cells. Two forms of the recombinant protein were detected, a monomer of 58 kDa and a dimer of 110 kDa, both recognized by p24-specific antibodies, indicating that the antigen was properly folded also in the homodimers. The purified fusion protein was used to immunize mice and elicit a specific antibody response, in a dose-dependent manner, that was detected in mice sera. Moreover, T-cell proliferation to p24 was stimulated in spleen cells of immunized animals, indicating that the fusion to immunoglobulin partners could represent a strategy to design new highly immunogenic plant-derived antigens (Obregon *et al.*, 2006).

In 2008 the production of p24, p24/p17 (truncated Gag), Pr55gag precursor protein, in *N. benthamiana* was described. The authors reported the comparisons between the expression levels of HIV-1 components in transgenic plants and using two different transient expression systems, one based on a tobacco mosaic virus (TMV) vector and the other on *Agrobacterium*-mediated gene transfer. Moreover, the influence of subcellular (chloroplast or ER) localization on protein yield was evaluated. The results showed that p24 and p17/p24 were expressed at higher levels compared to Pr55Gag in all systems and that the best yield was obtained with chloroplast-targeted proteins. Mice immunizations were performed using the p17/p24 partially purified protein, alone or as boost in animals primed by *gag* DNA vaccine. The results showed that the plant-derived viral protein was not immunogenic in mice per se but able to boost significantly both humoral and T-cell mediated specific immune responses in animals primed with *tat* DNA (Meyers *et al.*, 2008).

In the attempt to develop edible vaccines, mice oral immunization with fresh transgenic *Arabidopsis thaliana* tissues expressing p24 protein was described. Anti-Gag IgG were detected in sera only after injection with recombinant p37 (p17/p24) Gag, indicating that HIV protein contained in plant tissues is able to prime the anti-p24 immune response (Lindh *et al.*, 2008). In 2009, Scotti and colleagues published the first example of Pr55gag processing and assembly into VLPs in transgenic tobacco chloroplasts (Scotti *et al.*, 2009). Unfortunately, the immunogenic potential of plant-produced HIV VLPs was not reported.

As the transmembrane region of HIV-1 glycoprotein 41 (gp41), known as MPER domain, is critical in inducing a strong virus-specific neutralizing humoral immune response, Kessans and colleagues hypothesized that VLPs consisting of Gag and MPER could represent a broadly efficacious subunit HIV vaccine. The expression of these chimeric VLPs was achieved by transient and stable transformation in *N. benthamiana*. Biochemical characterization showed not only the formation of Gag VLPs but also the accumulation of chimeric VLPs including gp41 domain (Kessans *et al.*,

2013). Data concerning the immunogenicity of plant-produced HIV-1 VLPs are not yet available, but these particles could represent a promising strategy for the generation of anti-HIV vaccines.

9.2.3 Chimeric proteins carrying HIV envelope peptides

Viral pathogens like HIV-1 have developed the strategy to rapidly mutate surface proteins to escape recognition by immune cells. However, viral envelope proteins still harbour highly conserved peptides that are often associated with the production of neutralizing antibodies and known as 'protective epitopes'. Even though cellular immunity may provide control of HIV replication, neutralizing antibodies represent the first protective barrier useful to reduce the viral load during the early stage of infection. Studies with human monoclonal antibodies allowed the identification and localization of neutralizing epitopes in the two HIV-1 envelope glycoproteins, gp120 and gp41 (Zolla-Pazner, 2004; Burton *et al.*, 2012). As small molecules, like peptides, have low immunogenic potential, one strategy to produce subunit vaccines consists of the fusion of protective epitopes to heterologous carrier proteins (CPs). In this context, CPs of plant viruses have been exploited as effective peptide presentation systems. The majority of plant virus capsids is built by multiple copies of one or few structural proteins able to self-assemble into three-dimensional icosahedral or helical structures (Yildiz *et al.*, 2011). The highly ordered and repetitive surface renders these particles most effective in presenting antigens to the immune system, activating antibody responses (Jennings and Bachmann, 2008). For their characteristics, chimeric plant viruses (CVPs) have been successfully used as carriers of immunogenic peptides generating chimeric viral particles. The first study concerning the production of plant CVPs exposing the gp41 trans-membrane region spanning amino acid residues 731–752, was published in 1995 (McLain *et al.*, 1995). This peptide containing a neutralizing HIV-1 epitope was genetically fused to the cowpea mosaic virus (CPMV) CP and expressed in plants to produce CVPs. Purified chimeric virions injected subcutaneously in mice were able to stimulate the production of neutralizing antibodies. Concerning mucosal immunization, which is an ideal administration route to induce a protective immune response to virus dissemination, CPMV CVPs were able to produce IgA in faeces in mice immunized intranasally (Durrani *et al.*, 1998). Using the approach of fusing a highly conserved ELDKWA (2F5) epitope from gp41 to the N-terminus of potato virus X (PVX) CP, our laboratory produced CVPs in *N. benthamiana* tissues. In this work, the human immune response to CVPs carrying the 2F5 epitopes was studied in severe combined immunodeficient mice reconstituted with human peripheral blood lymphocytes (hu-PBL-SCID). Hu-PBL-SCID mice immunized with CVP-pulsed autologous dendritic cells were able to mount a specific human neutralizing antibody response against the gp41 epitope (Marusic *et al.*, 2001). A different strategy was used to produce the P1 gp41 peptide that includes the 2F5 epitope, in transgenic *N. benthamiana* plants. The sequence encoding the HIV-1 peptide was fused to the C-terminal region of the CTB subunit. Intranasal administration of purified chimeric oligomers were able to induce neutralizing responses and immunological memory in mice (Matoba *et al.*, 2004).

Recently, our research group described the production of artichoke mottled crinkle virus (AMCV)-derived chimeric VLPs exposing the 2F5 epitope on their surface. In this work, bioinformatics *in silico* predictions were useful in directing the experimental strategy used to fuse the 2F5 to the viral CP. The chimeric structural protein transiently expressed in *N. benthamiana* leaves was able to assemble into VLPs that were purified and characterized (Arcangeli *et al.*, 2013). Very few examples exist in literature concerning plant production of chimeric proteins carrying gp120 epitopes. In one of these studies the conserved V3 loop fused to CTB was expressed in potato with low yields (0.002–0.003% of TSP) (Kim *et al.*, 2004b). Moreover, an HIV-1 polypeptide containing V3 loop and C4 domain of gp120 was expressed in tobacco chloroplasts. Transgenic freeze-dried tobacco leaves used to immunize mice by the oral route were able to induce both systemic and mucosal antibody responses (Rubio-Infante *et al.*, 2012). Although CVPs demonstrated their efficacy in inducing neutralizing anti-HIV immune responses, VLPs which do not contain the viral genome and are not infectious represent ideal particles in terms of biosafety for the generation of novel plant virus-derived vaccine components.

9.3 Plant-made HIV Neutralizing Antibodies

9.3.1 Antibody expression in plants

Plants represent an advantageous expression system for the production of complex heterologous proteins, in particular immunotherapeutic reagents. There are many examples in literature of recombinant antibodies intended for both therapy and diagnosis that have been successfully expressed in plants (De Muynck *et al.*, 2010). Among these, monoclonal antibodies (mAbs) are efficiently folded and assembled within the ER of plant cells, retaining their full binding activity (Ma *et al.*, 1995). Antibody production in plants has been achieved using different expression strategies including stable transformation of the nuclear/chloroplast genome, or transient transformation using viral or *Agrobacterium*-based vectors (Garabagi *et al.*, 2012). While the establishment of transgenic plants requires long and laborious procedures with usually low IgG yields (Ko and Koprowski, 2005; Ma *et al.*, 2005), innovative transient expression systems allow the rapid high-yield accumulation (1–2 weeks) of recombinant protein in different plant species (tobacco, *N. benthamiana*, lettuce and tomato) (Vaquero *et al.*, 1999; Hull *et al.*, 2005; Negrouk *et al.*, 2005; Orzaez *et al.*, 2006). Major advantages over traditional expression systems based on mammalian cells are the adaptability and speed of these novel plant expression systems but also the reduced costs (Komarova *et al.*, 2010). Moreover, this transient expression technology permits the rapid scalability of the process, making it suitable for industrial-scale applications.

9.3.2 Comparison of plant-made antibodies to their mammalian counterparts

Among the drawbacks related to plant expression platforms, post-translational glycan modifications (which are different between plants and animals) have been long considered to be the major issue for some applications restricting the competitiveness of plants with manufacturing in mammalian expression systems.

Plant-derived mAbs differ from their mammalian counterpart mainly in their N-glycan composition. In fact, while core N-glycans are similar in plants and mammals, complex N-glycans show substantial differences. N-acetylneuraminic acid (NeuAc) residues are typical of mammalian complex N-glycans, while plants present complex-type oligosaccharides with β1,2-xylose and α1,3-fucose carbohydrate groups (Gomord *et al.*, 2010). This was regarded as a limitation to the use of plant-derived mAbs in immunotherapy, as it has been suggested that xylose and fucose moieties may cause immunogenic reactions in mammals (Bencúrová *et al.*, 2004; Jin *et al.*, 2008) or interfere with the binding to Fcγ receptors (Bardor *et al.*, 2003; Forthal *et al.*, 2010). This possible limitation has been successfully addressed by modifying the N-glycosylation profile of antibodies expressed in plant systems. Among the different strategies, retention of immuno-globulins in the endoplasmic reticulum (ER) using C-terminal KDEL tags proved to be successful, leading to the production of mAbs with high mannose-type N-glycans devoid of β1,2-xylose and α1,3-fucose residues (Petruccelli *et al.*, 2006; Floss *et al.*, 2009; Loos *et al.*, 2011). Another approach explored the use of plant glyco-engineering in which the inactivation of endogenous plant-specific glycosyltransferases and/or their complementation by heterologous human glycosyltransferases was used to alter glycosylation patterns of heterologous proteins. Fucose and xylose residues were successfully excluded using RNA interference in both duckweed and *N. benthamiana* (Cox *et al.*, 2006; Strasser *et al.*, 2008), or specific knockout lines of *A. thaliana* (Schähs *et al.*, 2007; Loos *et al.*, 2011). Another successful glyco-engineering method exploited the co-expression of specific human glycosyltransferases in plants, generating recombinant antibodies with β1,4-galactose or sialic acid residues (Fujiyama *et al.*, 2007; Strasser *et al.*, 2009; Vézina *et al.*, 2009; Castilho *et al.*, 2010). In conclusion, recent advances in plant glyco-engineering have permitted the production of human-like recombinant glycoproteins and plant expression systems are now considered versatile platforms for manufacturing monoclonal antibodies even with improved characteristics compared to their mammalian counterpart.

9.3.3 Anti-HIV antibody cocktails

In the last few years novel plant expression technologies have gained rising consideration in the production of monoclonal antibodies for the prevention of HIV transmission. In fact, anti-HIV antibody cocktails, in the absence of an efficacious vaccine and alternatively to current antiretroviral drug treatments based on small molecules, have shown to reduce mother to child transmission in non-human primates and are currently being formulated for use as vaginal microbicides and post-exposure prophylaxis/combination therapy. Monoclonal antibodies are advantageous compared to other molecules because they specifically inhibit HIV entry in human cells but can also induce the clearance of virus particles and infected cells by antibody-dependent cell toxicity and complement-dependent cytotoxicity (Armbruster *et al.*, 2004). There are currently several broadly neutralizing human mAbs available that hold promise as clinically therapeutic agents, which have been characterized either individually or as antibody cocktails in non-human primate studies and human clinical trials. Among these are: b12 (Kessler *et al.*, 1997) and VRC01 (Li *et al.*, 2011) targeting CD4 binding site; 2F5 (Conley *et al.*, 1994), 4E10 (Chen and Dierich, 1996) and m43 (Zhang *et al.*, 2012) directed against gp41 epitopes; and 2G12 recognizing a high mannose cluster in the C3-V4 region of gp120 (Trkola *et al.*, 1996). Three of these mAbs (2F5, 4E10 and 2G12), used as a cocktail, have recently completed a phase I clinical trial as vaginal microbicide candidate, showing no adverse effects on healthy women (Morris *et al.*, 2011.). However, HIV antibody cocktails in both therapy and prophylaxis may suffer from several drawbacks such as the high costs per treatment due to multiple doses required and the possible development of resistant viral strains. It must be noted that antibodies used as microbicides in prevention require administration of high doses, up to 5 mg per treatment twice a week with an estimate of 5000 kg of mAb required per year per 10 million women (Shattock and Moore, 2003). For the accessibility of mAb-based micro-bicides to wide population groups, especially in developing countries, a cost-effective large-scale production system would be required. Innovative plant expression systems, allowing large-scale production at reduced costs as well as adaptability and manufacturing speed, could meet these requirements favouring the use of mAbs, rather than small-molecule based antiretrovirals, in HIV prophylaxis.

9.3.4 Broadly neutralizing anti-HIV mAbs expressed in plants

Different studies have been published on the over-expression and characterization of plant-derived anti-HIV neutralizing antibodies (Table 9.2). Among the first anti-HIV mAbs to be expressed in plants, b12 produced in tobacco was characterized for its HIV-1 neutralization activity as an IgG and also a fusion with the anti-lectin cyanovirin-N (Sexton *et al.*, 2009). The gp41 binding mAb 4E10 has been successfully produced in hydroponic tobacco cultures as secretion in the culture medium (Drake *et al.*, 2009), in transgenic tobacco (Platis *et al.*, 2009) or by a transient agro-infiltration expression system in *N. benthamiana* plants (Rosenberg *et al.*, 2013). In particular two mAbs, the anti-gp120 2G12 and the anti-gp41 2F5, have been extensively characterized. In the case of the 2G12 antibody, directing the expression to the seed endosperm of transgenic maize gave very promising results with production levels of 40–100 mg kg^{-1} dry seed weight

(Rademacher *et al.*, 2008; Ramessar *et al.*, 2008). Moreover, both studies showed that the neutralizing activity of the plant-purified antibody was equivalent to or even better than that of the CHO (Chinese hamster ovary cells)-derived 2G12. Tobacco plants producing 2G12 have been obtained within the Pharma-Planta project (EU framework program 6). In 2011 this human monoclonal antibody (mAb) entered Europe's first clinical trial to test a pharmaceutical protein made in genetically modified tobacco plants (Fox, 2011). The same antibody was also transiently expressed in *N. benthamiana* by leaf agro-infiltration giving yields of 50 mg kg^{-1} of fresh biomass (Strasser *et al.*, 2008) or using a viral vector based on CPMV in *N. benthamiana* that yielded 100 mg kg^{-1} FW of mAb 2G12 (Sainsbury *et al.*, 2010). In the case of the mAb 2G12 expressed in transgenic tobacco plants it was shown that fusion of the antibody heavy chain (HC) and light chain (LC) to an elastin-like peptide (ELP) repeat raised production yields mainly by enhancing protein stability (Floss *et al.*, 2009). The mAb 2F5 has been expressed in transgenic tobacco cell suspension cultures yielding 6.4 mg kg^{-1} wet cell weight (Sack *et al.*, 2007). In this case, although the plant-derived 2F5 showed similar antigen binding activity compared to its CHO-derived counterpart, HIV-1 neutralization assays revealed a decreased efficiency. The mAb 2F5 has been also produced in transgenic maize seeds, but observed antibody yield was significantly lower than that reported for mAb 2G12 functional characterization of the recombinant HIV-neutralizing monoclonal antibody 2F5 produced in maize seeds (Sabalza *et al.*, 2012). The potential of using novel transient expression systems based on leaf infiltration of *Agrobacterium* for the production of anti-HIV mAb cocktails has been recently described by Rosenberg and colleagues (Rosenberg *et al.*, 2013). In this work, three different broadly neutralizing antibodies (b12, 2G12, and VRC01) were produced in *N. benthamiana* using leaf agro-infiltration and purified mAbs showed equivalent or better neutralization and antibody-dependent viral inhibition than their mammalian counterpart. Taken together these results demonstrate that novel plant expression systems could meet the requirements for a large-scale economical production of HIV-1 neutralizing antibodies.

9.4 Conclusions

The large number of papers published in the last two decades describing the expression of both vaccine components and neutralizing antibodies give evidence that plants represent a safe economical and scalable system for the production of biopharmaceuticals. In particular, transient expression technologies could offer new opportunities to developing countries, facilitating both development and production of bio-pharmaceuticals destined to the poor. As an example, in 2012 the Council for Scientific and Industrial Research (CSIR) in South Africa and a private company (ICON, Munich/Halle, Germany) announced an agreement that will allow the CSIR to make use of a plant-based manufacturing platform (magnICON®) for research and development and royalty-free manufacturing of rabies vaccines and post-exposure prophylaxis antibodies in the sub-Saharan region. In this context, the European 'Pharma-Planta' consortium represented the most important example of collaboration with developing countries aimed at the production of an anti-HIV neutralizing antibody for prophylaxis. This initiative, comprising academic and industry partners from Europe and South Africa, led to the development of a Good Manufacturing Practice (GMP)-compliant production process in plants that allowed successful completion of clinical trial phase I (http://www.pharma-planta.net). Novel economical and scalable GMP systems based on plant systems will allow in the near future other anti-HIV candidate molecules (antibodies/vaccines) to enter the pipeline.

References

Abela, I.A., Reynell, L. and Trkola, A. (2010) Therapeutic antibodies in HIV treatment – classical approaches to novel advances. *Current Pharmaceutical Des*ign 16, 3754–3766.

Addo, M.M., Yu, X.G., Rathod, A., Cohen, D., Eldridge, R.L., Strick, D., Johnston, M.N., Corcoran, C., Wurcel, A.G., Fitzpatrick, C.A., Feeney, M.E., Rodriguez, W.R., Basgoz, N., Draenert, R., Stone, D.R., Brander, C., Goulder, P.J., Rosenberg, E.S., Altfeld, M. and Walker, B.D. (2003) Comprehensive epitope analysis of human immunodeficiency virus type 1 (HIV-1)-specific T-cell responses directed against the entire expressed HIV-1 genome demonstrate broadly directed responses, but no correlation to viral load. *Journal of Virology* 77, 2081–2092.

Arcangeli, C., Circelli, P., Donini, M., Aljabali, A.A., Benvenuto, E., Lomonossoff, G.P. and Marusic, C. (2013) Structure-based design and experimental engineering of a plant virus nanoparticle for the presentation of immunogenic epitopes and as a drug carrier. *Journal of Biomolecular and Structural Dynamics* 32(4), 630–647.

Armbruster, C., Stiegler, G.M., Vcelar, B.A., Jäger, W., Köller, U., Jilch, R., Ammann, C.G., Pruenster, M., Stoiber, H. and Katinger, H.W.J. (2004) Passive immunization with the anti-HIV-1 human monoclonal antibody (hMAb) 4E10 and the hMAb combination 4E10/2F5/2G12. *Journal of Antimicrobial Chemotherapy* 54, 915–920.

Azad, A.A., Failla, P., Lucantoni, A., Bentley, J., Mardon, C., Wolfe, A., Fuller, K., Hewish, D., Sengupta, S., Sankovich, S., Grgacic, E., McPhee, D. and Macreadie, I. (1994) Large-scale production and characterization of recombinant human immunodeficiency virus type 1 Nef. *Journal of General Virology* 75, 651–655.

Barbante, A., Irons, S., Hawes, C., Frigerio, L., Vitale, A. and Pedrazzini, E. (2008) Anchorage to the cytosolic face of the ER membrane: a new strategy to stabilize a cytosolic recombinant antigen in plants. *Plant Biotechnology Journal* 6, 560–575.

Bardor, M., Faveeuw, C., Fitchette, A.C., Gilbert, D., Galas, L., Trottein, F., Faye, L. and Lerouge, P. (2003) Immunoreactivity in mammals of two typical plant glyco-epitopes, core alpha (1,3)-fucose and core xylose. *Glycobiology* 13, 427–434.

Beachy, R.N., Fitchen, J.H. and Hein, M.B. (1996) Use of plant viruses for delivery of vaccine epitopes. *Annals of the New York Academy of Sciences* 792, 43–49.

Bencúrová, M., Hemmer, W., Focke-Tejkl, M., Wilson, I.B. and Altmann, F. (2004) Specificity of IgG and IgE antibodies against plant and insect glycoprotein glycans determined with artificial glycoforms of human transferrin. *Glycobiology* 14, 457–466.

Burton, D.R., Poignard, P., Stanfield, R.L. and Wilson, I.A. (2012) Broadly neutralizing antibodies present new prospects to counter highly antigenically diverse viruses. *Science* 337, 183–186.

Cafaro, A., Caputo, A., Fracasso, C., Maggiorella, M.T., Goletti, D., Baroncelli, S., Pace, M., Sernicola, L., Koanga-Mogtomo, M.L., Betti, M., Borsetti, A., Belli, R., Akerblom, L., Corrias, F., Buttò, S., Heeney, J., Verani, P., Titti, F. and Ensoli, B. (1999) Control of SHIV-89.6P-infection of cynomolgus monkeys by HIV-1 Tat protein vaccine. *Nature Medicine* 5, 643–650.

Castilho, A., Strasser, R., Stadlmann, J., Grass, J., Jez, J., Gattinger, P., Kunert, R., Quendler, H., Pabst, M., Leonard, R., Altmann, F. and Steinkellner, H. (2010) In planta protein sialylation through overexpression of the respective mammalian pathway. *Journal of Biological Chemistry* 285, 15923–15930.

Chen, Y.H. and Dierich, M.P. (1996) Identification of a second site in HIV-1 gp41 mediating binding to cells. *Immunology Letters* 52, 153–156.

Collins, K.L., Chen, B.K., Kalams, S.A., Walker, B.D. and Baltimore, D. (1998) HIV-1 Nef protein protects infected primary cells against killing by cytotoxic T lymphocytes. *Nature* 391, 397–401.

Conley, A.J., Kessler, J.A. 2nd, Boots, L.J., Tung, J.S., Arnold, B.A., Keller, P.M., Shaw, A.R. and Emini, E.A. (1994) Neutralization of divergent human immunodeficiency virus type 1 variants and primary isolates by IAM-41-2F5, an anti-gp41 human monoclonal antibody. *Proceedings of the National Academy of Sciences USA* 91, 3348–3352.

Cooke, S.J., Coates, K., Barton, C.H., Biggs, T.E., Barrett, S.J., Cochrane, A., Oliver, K., McKeating, J.A., Harris, M.P. and Mann, D.A. (1997) Regulated expression vectors demonstrate cell-type-specific sensitivity to human immunodeficiency virus type 1 Nef-induced cytostasis. *Journal of General Virology* 78, 381–392.

Cox, K.M., Sterling, J.D., Regan, J.T., Gasdaska, J.R., Frantz, K.K., Peele, C.G., Black, A., Passmore, D., Moldovan-Loomis, C., Srinivasan, M., Cuison, S., Cardarelli, P.M. and Dickey, L.F. (2006) Glycan optimization of a human monoclonal antibody in the aquatic plant *Lemna minor*. *Nature Biotechnology* 24, 1591–1597.

Cruz, N.V., Amorim, R., Oliveira, F.E., Speranza, F.A. and Costa, L.J. (2013) Mutations in the nef and vif genes associated with progression to AIDS in elite controller and slow-progressor patients. *Journal of Medical Virology* 85(4), 563–574.

Cueno, M.E., Hibi, Y., Karamatsu, K., Yasutomi, Y., Imai, K., Laurena, A.C. and Okamoto, T. (2010) Preferential expression and immunogenicity of HIV-1 Tat fusion protein expressed in tomato plant. *Transgenic Research* 19, 889–895.

De Muynck, B., Navarre, C. and Boutry, M. (2010) Production of antibodies in plants: status after twenty years. *Plant Biotechnology Journal* 8, 529–563.

de Virgilio, M., De Marchis, F., Bellucci, M., Mainieri, D., Rossi, M., Benvenuto, E., Arcioni, S. and Vitale, A. (2008) The human immunodeficiency virus antigen Nef forms protein bodies in leaves of transgenic tobacco when fused to zeolin. *Journal of Experimental Botany* 59, 2815–2829.

Debaisieux, S., Rayne, F., Yezid, H. and Beaumelle, B. (2012) The ins and outs of HIV-1 Tat. *Traffic* 13, 355–363.

Doan, L.X., Li, M., Chen, C. and Yao, Q. (2005) Virus-like particles as HIV-1 vaccines. *Reviews in Medical Virology* 15, 75–88.

Drake, P.M., Barbi, T., Sexton, A., McGowan, E., Stadlmann, J., Navarre, C., Paul, M.J. and Ma, J.K. (2009) Development of rhizosecretion as a production system for recombinant proteins from hydroponic cultivated tobacco. *The FASEB Journal* 23, 3581–3589.

Durrani, Z., McInerney, T.L., McLain, L., Jones, T., Bellaby, T., Brennan, F.R. and Dimmock, N.J. (1998) Intranasal immunization with a plant virus expressing a peptide from HIV-1 gp41 stimulates better mucosal and systemic HIV-1-specific IgA and IgG than oral immunization. *Journal of Immunological Methods* 220, 93–103.

Floss, D.M., Sack, M., Stadlmann, J., Rademacher, T., Scheller, J., Stöger, E., Fischer, R. and Conrad, U. (2008) Biochemical and functional characterization of anti-HIV antibody-ELP fusion proteins from transgenic plants. *Plant Biotechnology Journal* 6, 379–391.

Floss, D.M., Sack, M., Arcalis, E., Stadlmann, J., Quendler, H., Rademacher, T., Stoger, E., Scheller, J., Fischer, R. and Conrad, U. (2009) Influence of elastin-like peptide fusions on the quantity and quality of a tobacco-derived human immunodeficiency virus-neutralizing antibody. *Plant Biotechnology Journal* 7, 899–913.

Forthal, D.N., Gach, J.S., Landucci, G., Jez, J., Strasser, R., Kunert, R. and Steinkellner, H. (2010) Fc-glycosylation influences Fcγ receptor binding and cell-mediated anti-HIV activity of monoclonal antibody 2G12. *Journal of Immunology* 185, 6876–6882.

Fox, J.L. (2011) HIV drugs made in tobacco. *Nature Biotechnology* 29, 852.

Fujiyama, K., Furukawa, A., Katsura, A., Misaki, R., Omasa, T. and Seki, T. (2007) Production of mouse monoclonal antibody with galactose-extended sugar chain by suspension cultured tobacco BY2 cells expressing human beta(1,4)-galactosyltransferase. *Biochemical and Biophysical Research Communications* 358, 85–91.

Garabagi, F., McLean, M.D. and Hall, J.C. (2012) Transient and stable expression of antibodies in *Nicotiana* species. *Methods in Molecular Biology* 907, 389–408.

Geyer, M., Fackler, O.T. and Peterlin, B.M. (2001) Structure-function relationships in HIV-1 Nef. *EMBO Reports* 2, 580–585.

Gomord, V., Fitchette, A.C., Menu-Bouaouiche, L., Saint-Jore-Dupas, C., Plasson, C., Michaud, D. and Faye, L. (2010) Plant-specific glycosylation patterns in the context of therapeutic protein production. *Plant Biotechnology Journal* 8, 564–587.

Gonzalez-Rabade, N., McGowan, E.G., Zhou, F., McCabe, M.S., Bock, R., Dix, P.J., Gray, J.C. and Ma, J.K. (2011) Immunogenicity of chloroplast-derived HIV-1 p24 and a p24-Nef fusion protein following subcutaneous and oral administration in mice. *Plant Biotechnology Journal* 9, 629–638.

Hamorsky, K.T., Grooms-Williams, T.W., Husk, A.S., Bennett, L.J., Palmer, K.E. and Matoba, N. (2013) Efficient single tobamoviral vector-based bioproduction of broadly neutralizing anti-HIV-1 monoclonal antibody VRC01 in *Nicotiana benthamiana* plants and utility of VRC01 in combination microbicides. *Antimicrobial Agents and Chemotherapy* 57, 2076–2086.

Holland, T., Sack, M., Rademacher, T., Schmale, K., Altmann, F., Stadlmann, J., Fischer, R. and Hellwig, S. (2010) Optimal nitrogen supply as a key to increased and sustained production of a monoclonal full-size antibody in BY-2 suspension culture. *Biotechnology and Bioengineering* 107, 278–289.

Hull, A.K., Criscuolo, C.J., Mett, V., Groen, H., Steeman, W., Westra, H., Chapman, G., Legutki, B., Baillie, L. and Yusibov, V. (2005) Human-derived, plant-produced monoclonal antibody for the treatment of anthrax. *Vaccine* 23, 2082–2086.

Jennings, G.T. and Bachmann, M.F. (2008) The coming of age of virus-like particle vaccines. *Biological Chemistry* 389, 521–536.

Jin, C., Altmann, F., Strasser, R., Mach, L., Schähs, M., Kunert, R., Rademacher, T., Glössl, J. and Steinkellner, H. (2008) A plant-derived human monoclonal antibody induces an anti-carbohydrate immune response in rabbits. *Glycobiology* 18, 235–241.

Joelson, T., Akerblom, L., Oxelfelt, P., Strandberg, B., Tomenius, K. and Morris, T.J. (1997) Presentation of a foreign peptide on the surface of tomato bushy stunt virus. *Journal of General Virology* 78, 1213–1217.

Karasev, A.V., Foulke, S., Wellens, C., Rich, A., Shon, K.J., Zwierzynski, I., Hone, D., Koprowski, H. and Reitz, M. (2005) Plant based HIV-1 vaccine candidate: Tat protein produced in spinach. *Vaccine* 23, 1875–1880.

Kessans, S.A., Linhart, M.D., Matoba, N. and Mor, T. (2013) Biological and biochemical characterization of HIV-1 Gag/dgp41 virus-like particles expressed in *Nicotiana benthamiana*. *Plant Biotechnology Journal* 11(6), 681–690.

Kessler, J.A. 2nd, McKenna, P.M., Emini, E.A., Chan, C.P., Patel, M.D., Gupta, S.K., Mark, G.E. 3rd, Barbas, C.F. 3rd, Burton, D.R. and Conley, A.J. (1997) Recombinant human monoclonal antibody IgG1b12 neutralizes diverse human immunodeficiency virus type 1 primary isolates. *AIDS Research and Human Retroviruses* 13, 575–582.

Kim, T.G., Ruprecht, R. and Langridge, W.H. (2004a) Synthesis and assembly of a cholera toxin B subunit SHIV 89.6p Tat fusion protein in transgenic potato. *Protein Expression and Purification* 35, 313–319.

Kim, T.G., Gruber, A. and Langridge, W.H. (2004b) HIV-1 gp120 V3 cholera toxin B subunit fusion gene expression in transgenic potato. *Protein Expression and Purification* 37, 196–202.

Ko, K. and Koprowski, H. (2005) Plant biopharming of monoclonal antibodies. *Virus Research* 111, 93–100.

Kohleisen, B., Gaedigk-Nitschko, K., Ohlmann, M., Götz, E., Ostolaza, H., Goni, F.M. and Erfle, V. (1996) Heparin-binding capacity of the HIV-1 NEF-protein allows one-step purification and biochemical characterization. *Journal of Virological Methods* 60, 89–101.

Komarova, T.V., Baschieri, S., Donini, M., Marusic, C., Benvenuto, E. and Dorokhov, Y.L. (2010) Transient expression systems for plant-derived biopharmaceuticals. *Expert Review of Vaccines* 9, 859–876.

Li, Y., O'Dell, S., Walker, L.M., Wu, X., Guenaga, J., Feng, Y., Schmidt, S.D., McKee, K., Louder, M.K., Ledgerwood, J.E., Graham, B.S., Haynes, B.F., Burton, D.R., Wyatt, R.T. and Mascola, J.R. (2011) Mechanism of neutralization by the broadly neutralizing HIV-1 monoclonal antibody VRC01. *Journal of Virology* 85, 8954–8967.

Lindh, I., Kalbina, I., Thulin, S., Scherbak, N., Sävenstrand, H., Bråve, A., Hinkula, J., Strid, A. and Andersson, S. (2008) Feeding of mice with *Arabidopsis thaliana* expressing the HIV-1 subtype C p24 antigen gives rise to systemic immune responses. *Acta Pathologica, Microbiologica, et Immunologica Scandinavica* 116, 985–994.

Lombardi, R., Circelli, P., Villani, M.E., Buriani, G., Nardi, L., Coppola, V., Bianco, L., Benvenuto, E., Donini, M. and Marusic, C. (2009) High-level HIV-1 Nef transient expression in *Nicotiana benthamiana* using the P19 gene silencing suppressor protein of Artichoke Mottled Crinckle Virus. *BMC Biotechnology* 9, e96.

Loos, A., Van Droogenbroeck, B., Hillmer, S., Grass, J., Kunert, R., Cao, J., Robinson, D.G., Depicker, A. and Steinkellner, H. (2011) Production of monoclonal antibodies with a controlled N-glycosylation pattern in seeds of *Arabidopsis thaliana*. *Plant Biotechnology Journal* 9, 179–192.

Ma, J.K., Hiatt, A., Hein, M., Vine, N.D., Wang, F., Stabila, P., van Dolleweerd, C., Mostov, K. and Lehner, T. (1995) Generation and assembly of secretory antibodies in plants. *Science* 268, 716–719.

Ma, J.K., Drake, P.M., Chargelegue, D., Obregon, P. and Prada, A. (2005) Antibody processing and engineering in plants, and new strategies for vaccine production. *Vaccine* 23, 1814–1818.

Macreadie, I.G., Fernley, R., Castelli, L.A., Lucantoni, A., White, J. and Azad, A. (1998) Expression of HIV-1 nef in yeast causes membrane perturbation and release of the myristylated Nef protein. *Journal of Biomedical Science* 5, 203–210.

Marusic, C., Rizza, P., Lattanzi, L., Mancini, C., Spada, M., Belardelli, F., Benvenuto, E. and Capone, I. (2001) Chimeric plant virus particles as immunogens for inducing murine and human immune responses against human immunodeficiency virus type 1. *Journal of Virology* 75, 8434–8439.

Marusic, C., Nuttall, J., Buriani, G., Lico, C., Lombardi, R., Baschieri, S., Benvenuto, E. and Frigerio, L. (2007) Expression, intracellular targeting and purification of HIV Nef variants in tobacco cells. *BMC Biotechnology* 7, e12.

Marusic, C., Vitale, A., Pedrazzini, E., Donini, M., Frigerio, L., Bock, R., Dix, P.J., McCabe, M.S., Bellucci, M. and Benvenuto, E. (2009) Plant-based strategies aimed at expressing HIV antigens and neutralizing antibodies at high levels. Nef as a case study. *Transgenic Research* 18, 499–512.

Matoba, N., Magérus, A., Geyer, B.C., Zhang, Y., Muralidharan, M., Alfsen, A., Arntzen, C.J., Bomsel, M. and Mor, T.S. (2004) A mucosally targeted subunit vaccine candidate eliciting HIV-1 transcytosis-blocking Abs. *Proceedings of the National Academy of Sciences USA* 101, 13584–13589.

Matoba, N., Kajiura, H., Cherni, I., Doran, J.D., Bomsel, M., Fujiyama, K. and Mor, T.S. (2009) Biochemical and immunological characterization of the plant-derived candidate human immunodeficiency virus type 1 mucosal vaccine CTB-MPR. *Plant Biotechnology Journal* 7, 129–145.

Maxmen, A. (2012) Drug-making plant blooms. *Nature* 485, 160.

McCabe, M.S., Klaas, M., Gonzalez-Rabade, N., Poage, M., Badillo-Corona, J., Zhou, F., Karcher, D., Bock, R. and Gray, J.C. (2008) Plastid transformation of high biomass tobacco variety Maryland Mammoth for production of HIV-1 p24 antigen. *Plant Biotechnology Journal* 6, 914–929.

McInerney, T.L., Brennan, F.R., Jones, T.D. and Dimmock, N.J. (1999) Analysis of the ability of five adjuvants to enhance immune responses to a chimeric plant virus displaying an HIV-1 peptide. *Vaccine* 17, 1359–1368.

McLain, L., Porta, C., Lomonossoff, G.P., Durrani, Z. and Dimmock, N.J. (1995) Human immunodeficiency virus type 1-neutralizing antibodies raised to a glycoprotein 41 peptide expressed on the surface of a plant virus. *AIDS Research and Human Retroviruses* 11, 327–334.

McLain, L., Durrani, Z., Wisniewski, L.A., Porta, C., Lomonossoff, G.P. and Dimmock, N.J. (1996) Stimulation of neutralizing antibodies to human immunodeficiency virus type 1 in three strains of mice immunized with a 22 amino acid peptide of gp41 expressed on the surface of a plant virus. *Vaccine* 14, 799–810.

Meyers, A., Chakauya, E., Shephard, E., Tanzer, F.L., Maclean, J., Lynch, A., Williamson, A.L. and Rybicki, E.P. (2008) Expression of HIV-1 antigens in plants as potential subunit vaccines. *BMC Biotechnology* 8, 53.

Morris, G., Wiggins, R., Woodhall, S., Taylor, C., Vcelar, B., Bland, B. and Lacey, C. (2011) A phase I randomized controlled trial of a triple anti-HIV-1 monoclonal antibody vaginal microbicide. Paper no 990. *18th Conference on Retroviruses and Opportunistic Infections*, Boston, Massachusetts.

Negrouk, V., Eisner, G., Lee, H., Han, K., Taylor, D. and Wong, H.C. (2005) Highly efficient transient expression of functional recombinant antibodies in lettuce. *Plant Science* 169, 433–438.

Novitsky, V., Cao, H., Rybak, N., Gilbert, P., McLane, M.F., Gaolekwe, S., Peter, T., Thior, I., Ndung'u, T., Marlink, R., Lee, T.H. and Essex, M. (2002) Magnitude and frequency of cytotoxic T-lymphocyte responses: identification of immunodominant regions of human immunodeficiency virus type 1 subtype C. *Journal of Virology* 76, 10155–10168.

Obregon, P., Chargelegue, D., Drake, P.M., Prada, A., Nuttall, J., Frigerio, L. and Ma, J.K. (2006) HIV-1 p24-immunoglobulin fusion molecule: a new strategy for plant-based protein production. *Plant Biotechnology Journal* 4, 195–207.

Orzaez, D., Mirabel, S., Wieland, W.H. and Granell, A. (2006) Agroinjection of tomato fruits. A tool for rapid functional analysis of transgenes directly in fruit. *Plant Physiology* 140, 3–11.

Pena-Ramírez, Y.J., Tasciotti, E., Gutierrez-Ortega, A., Donayre Torres, A.J., Olivera Flores, M.T., Giacca, M. and Gómez Lim, M.A. (2007) Fruit-specific expression of the human immunodeficiency virus type 1 tat gene in tomato plants and its immunogenic potential in mice. *Clinical and Vaccine Immunology* 6, 685–692.

Peng, B. and Robert-Guroff, M. (2001) Deletion of N-terminal myristoylation site of HIV Nef abrogates both MHC-1 and CD4 down-regulation. *Immunology Letters* 78, 195–200.

Peng, B., Voltan, R., Cristillo, A.D., Alvord, W.G., Davis-Warren, A., Zhou, Q., Murthy, K.K. and Robert-Guroff, M. (2006) Replicating Ad-recombinants encoding non-myristoylated rather than wild-type HIV Nef elicit enhanced cellular immunity. *AIDS* 20, 2149–2157.

Pérez-Filgueira, D.M., Brayfield, B.P., Phiri, S., Borca, M.V., Wood, C. and Morris, T.J. (2004) Preserved antigenicity of HIV-1 p24 produced and purified in high yields from plants inoculated with a tobacco mosaic virus (TMV)-derived vector. *Journal of Virological Methods* 121, 201–208.

Petruccelli, S., Otegui, M.S., Lareu, F., Tran Dinh, O., Fitchette, A.C., Circosta, A., Rumbo, M., Bardor, M., Carcamo, R., Gomord, V. and Beachy, R.N. (2006) A KDEL-tagged monoclonal antibody is efficiently retained in the endoplasmic reticulum in leaves, but is both partially secreted and sorted to protein storage vacuoles in seeds. *Plant Biotechnology Journal* 4, 511–527.

Platis, D., Maltezos, A., Ma, J.K. and Labrou, N.E. (2009) Combinatorial de novo design and application of a biomimetic affinity ligand for the purification of human anti-HIV mAb 4E10 from transgenic tobacco. *Journal of Molecular Recognition* 22, 415–424.

Porta, C., Spall, V.E., Loveland, J., Johnson, J.E., Barker, P.J. and Lomonossoff, G.P. (1994) Development of cowpea mosaic virus as a high-yielding system for the presentation of foreign peptides. *Virology* 202, 949–955.

Rademacher, T., Sack, M., Arcalis, E., Stadlmann, J., Balzer, S., Altmann, F., Quendler, H., Stiegler, G., Kunert, R., Fischer, R. and Stoger, E. (2008) Recombinant antibody 2G12 produced in maize endosperm efficiently neutralizes HIV-1 and contains predominantly single-GlcNAc N-glycans. *Plant Biotechnology Journal* 6, 189–201.

Ramessar, K., Rademacher, T., Sack, M., Stadlmann, J., Platis, D., Stiegler, G., Labrou, N., Altmann, F., Ma, J., Stöger, E., Capell, T. and Christou, P. (2008) Cost-effective production of a vaginal protein microbicide to prevent HIV transmission. *Proceedings of the National Academy of Sciences USA* 105, 3727–3732.

Ramírez, Y.J., Tasciotti, E., Gutierrez-Ortega, A., Donayre Torres, A.J., Olivera Flores, M.T., Giacca, M. and Gómez Lim, M.A. (2007) Fruit-specific expression of the human immunodeficiency virus type 1 tat gene in tomato plants and its immunogenic potential in mice. *Clinical and Vaccine Immunology* 14, 685–692.

Rosales-Mendoza, S., Rubio-Infante, N., Govea-Alonso, D.O. and Moreno-Fierros, L. (2012) Current status and perspectives of plant-based candidate vaccines against the human immunodeficiency virus (HIV). *Plant Cell Reports* 31, 495–511.

Rosenberg, Y., Sack, M., Montefiori, D., Forthal, D., Mao, L., Hernandez-Abanto, S., Urban, L., Landucci, G., Fischer, R. and Jiang, X. (2013) Rapid high-level production of functional HIV broadly neutralizing monoclonal antibodies in transient plant expression systems. *PLoS One* 8, e58724.

Rubio-Infante, N., Govea-Alonso, D.O., Alpuche-Solís, Á.G., García-Hernández, A.L., Soria-Guerra, R.E., Paz-Maldonado, L.M., Ilhuicatzi-Alvarado, D., Varona-Santos, J.T., Verdín-Terán, L., Korban, S.S., Moreno-Fierros, L. and Rosales-Mendoza, S. (2012) A chloroplast-derived C4V3 polypeptide from the human immunodeficiency virus (HIV) is orally immunogenic in mice. *Plant Molecular Biology* 78, 337–349.

Sabalza, M., Madeira, L., van Dolleweerd, C., Ma, J.K., Capell, T. and Christou, P. (2012) Functional characterization of the recombinant HIV-neutralizing monoclonal antibody 2F5 produced in maize seeds. *Plant Molecular Biology* 80, 477–488.

Sack, M., Paetz, A., Kunert, R., Bomble, M., Hesse, F., Stiegler, G., Fischer, R., Katinger, H., Stoeger, E. and Rademacher, T. (2007) Functional analysis of the broadly neutralizing human anti-HIV-1 antibody 2F5 produced in transgenic BY-2 suspension cultures. *The FASEB Journal* 21, 1655–1664.

Sainsbury, F., Sack, M., Stadlmann, J., Quendler, H., Fischer, R. and Lomonossoff, G.P. (2010) Rapid transient production in plants by replicating and non-replicating vectors yields high quality functional anti-HIV antibody. *PLoS One* 5, e13976.

Schähs, M., Strasser, R., Stadlmann, J., Kunert, R., Rademacher, T. and Steinkellner, H. (2007) Production of a monoclonal antibody in plants with a humanized N-glycosylation pattern. *Plant Biotechnology Journal* 5, 657–663.

Scotti, N., Alagna, F., Ferraiolo, E., Formisano, G., Sannino, L., Buonaguro, L., De Stradis, A., Vitale, A., Monti, L., Grillo, S., Buonaguro, F.M. and Cardi, T. (2009) High-level expression of the HIV-1 Pr55gag polyprotein in transgenic tobacco chloroplasts. *Planta* 229, 1109–1122.

Scotti, N., Buonaguro, L., Tornesello, M.L., Cardi, T. and Buonaguro, F.M. (2010) Plant-based anti-HIV-1 strategies: vaccine molecules and antiviral approaches. *Expert Review of Vaccines* 9, 925–936.

Sexton, A., Harman, S., Shattock, R.J. and Ma, J.K. (2009) Design, expression, and characterization of a multivalent, combination HIV microbicide. *The FASEB Journal* 23, 3590–3600.

Shattock, R.J. and Moore, J.P. (2003) Inhibiting sexual transmission of HIV-1 infection. *Nature Reviews Microbiology* 1, 25–34.

Sirén, N., Weegar, J., Dahlbacka, J., Kalkkinen, N., Fagervik, K., Leisola, M. and vonWeymarn, N. (2006) Production of recombinant HIV-1 Nef (negative factor) protein using *Pichia pastoris* and a low-temperature fed-batch strategy. *Biotechnology and Applied Biochemistry* 44, 151–158.

Strasser, R., Stadlmann, J., Schähs, M., Stiegler, G., Quendler, H., Mach, L., Glössl, J., Weterings, K., Pabst, M. and Steinkellner, H. (2008) Generation of glyco-engineered *Nicotiana benthamiana* for the production of monoclonal antibodies with a homogeneous human-like N-glycan structure. *Plant Biotechnology Journal*, 6, 392–402.

Strasser, R., Castilho, A., Stadlmann, J., Kunert, R., Quendler, H., Gattinger, P., Jez, J., Rademacher, T., Altmann, F., Mach, L. and Steinkellner, H. (2009) Improved virus neutralization by plant-produced anti-HIV antibodies with a homogeneous b1,4-galactosylated N-glycan profile. *Journal of Biological Chemistry* 284, 20479–20485.

Sugiyama, Y., Hamamoto, H., Takemoto, S., Watanabe, Y. and Okada, Y. (1995) Systemic production of foreign peptides on the particle surface of tobacco mosaic virus. *FEBS Letters* 359, 247–250.

Trkola, A., Purtscher, M., Muster, T., Ballaun, C., Buchacher, A., Sullivan, N., Srinivasan, K., Sodroski, J., Moore, J.P. and Katinger, H. (1996) Human monoclonal antibody 2G12 defines a distinctive neutralization epitope on the gp120 glycoprotein of human immunodeficiency virus type 1. *Journal of Virology* 70, 1100–1108.

Vaquero, C., Sack, M., Chandler, J., Drossard, J., Schuster, F., Monecke, M., Schillberg, S. and Fischer, R. (1999) Transient expression of a tumor-specific single-chain fragment and a chimeric antibody in tobacco leaves. *Proceedings of the National Academy of Sciences USA* 96, 11128–11133.

Vézina, L.P., Faye, L., Lerouge, P., D'Aoust, M.A., Marquet-Blouin, E., Burel, C., Lavoie, P.O., Bardor, M. and Gomord, V. (2009) Transient co-expression for fast and high-yield production of antibodies with human-like N-glycans in plants. *Plant Biotechnology Journal* 7, 442–455.

Webster, D.E., Thomas, M.C., Pickering, R., Whyte, A., Dry, I.B., Gorry, P.R. and Wesselingh, S.L. (2005) Is there a role for plant-made vaccines in the prevention of HIV/AIDS? *Immunology & Cell Biology* 83, 239–247.

Yildiz, I., Shukla, S. and Steinmetz, N.F. (2011) Applications of viral nanoparticles in medicine. *Current Opinion in Biotechnology* 22, 901–908.

Yusibov, V., Modelska, A., Steplewski, K., Agadjanyan, M., Weiner, D., Hooper, D.C. and Koprowski, H. (1997) Antigens produced in plants by infection with chimeric plant viruses immunize against rabies virus and HIV-1. *Proceedings of the National Academy of Sciences USA* 94, 5784–5788.

Zhang, G., Leung, C., Murdin, L., Rovinski, B. and White, K.A. (2000) In planta expression of HIV-1 p24 protein using an RNA plant virus-based expression vector. *Molecular Biotechnology* 14, 99–107.

Zhang, G.G., Rodrigues, L., Rovinski, B. and White, K.A. (2002) Production of HIV-1 p24 protein in transgenic tobacco plants. *Molecular Biotechnology* 20, 131–136.

Zhang, M.Y., Yuan, T., Li, J., Rosa Borges, A., Watkins, J.D., Guenaga, J., Yang, Z., Wang, Y., Wilson, R., Li, Y., Polonis, V.R., Pincus, S.H., Ruprecht, R.M. and Dimitrov, D.S. (2012) Identification and characterization of a broadly cross-reactive HIV-1 human monoclonal antibody that binds to both gp120 and gp41. *PLoS One* 7, e44241.

Zhou, F., Badillo-Corona, J.A., Karcher, D., Gonzalez-Rabade, N., Piepenburg, K., Borchers, A.-M.I., Maloney, A.P., Kavanagh, T.A., Gray, J.C. and Bock, R. (2008) High-level expression of HIV antigens from the tobacco and tomato plastid genomes. *Plant Biotechnology Journal* 6, 897–913.

Zolla-Pazner, S. (2004) Identifying epitopes of HIV-1 that induce protective antibodies. *Nature Reviews Immunology* 4, 199–210.

Case Study 3: The Search for a Plant-made Vaccine for Pandemic Influenza Virus

Kathleen L. Hefferon*

Cornell University, Ithaca, New York

This book has largely dealt with selected examples of plant-made vaccines against infectious diseases that continue to be detrimental to developing countries. However, other infectious agents also exist that are considered to be a substantial burden both to highly industrialized nations and to the developing world. This last chapter briefly highlights some of the progress made with respect to the development of a plant-made vaccine to combat pandemic influenza virus. Pandemic flu not only causes substantial illness and death all over the world, but also creates a significant economic burden for those countries that are afflicted due to the high number of individuals who must leave the workforce for extended periods of time due to illness. This chapter concludes with a discussion of some of the events necessary and obstacles to be considered to ensure that plant-derived vaccines and other pharmaceuticals make a difference for those in need and become a reality for the developing world.

10.1 Plant-made Vaccines to Protect against Pandemic Influenza Virus

Influenza, commonly referred to as 'the flu', belongs to the negative-stranded group of RNA viruses and is responsible for high levels of mortality and morbidity throughout human populations worldwide. Highly contagious, it is believed that influenza could be responsible for 300,000 to 500,000 deaths annually, and as many as 5 million hospitalizations (WHO, 2010) (Fig. 10.1). The virus encodes the highly immunogenic haemagglutinin (HA) and neuraminidase (NA) surface glycoproteins; it is the frequent changes in the sequence of these surface proteins that give rise to much of the variation arising between different strains. Due to the segmented nature of the influenza genome, the genes that encode these two proteins undergo rapid re-assortment over the course of host co-infection, and result in new pathogenic strains on a yearly basis (Brammer *et al.*, 2011).

---

* E-mail: klh22@cornell.edu

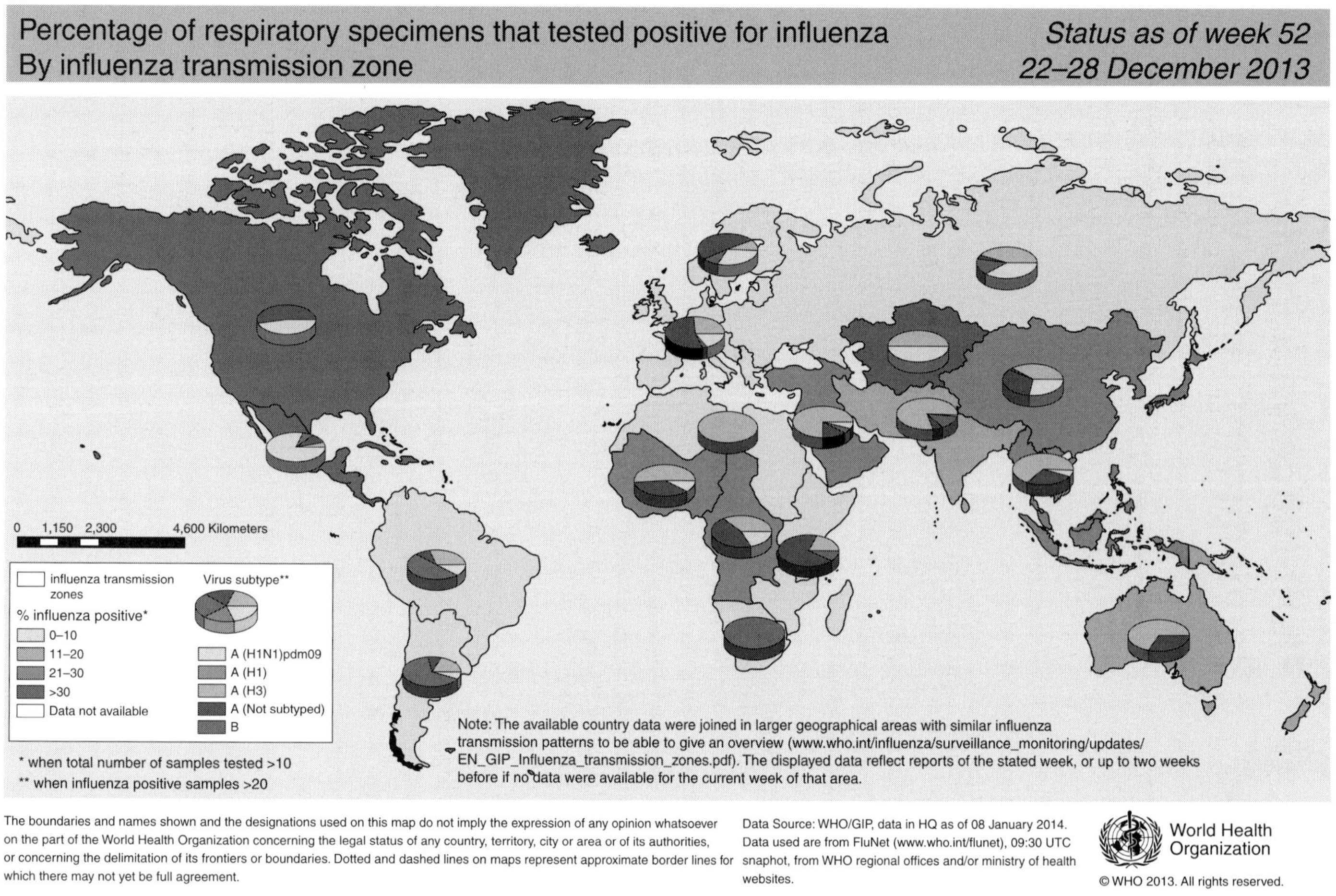

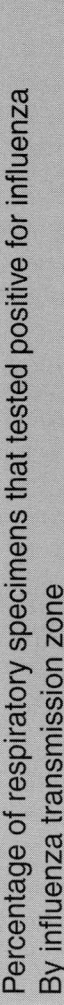

Fig. 10.1. Percentage of respiratory specimens that tested positive for influenza by influenza transmission zone (World Health Organization, 2013).

Due to the frequent changes in the sequence structure of these major surface glycoproteins, there is a universal need for a technology that enables new strains of the vaccine to be manufactured for every new flu season in sufficient quantities. However, a number of factors limit global pandemic influenza vaccine coverage, including the ability to rapidly respond to every new isolate that arises each flu season, as well as the problem of providing for sufficient quantities of influenza antigen to facilitate large-scale vaccine production (Perdue et al., 2011). In the past, the vast majority of widely used influenza vaccines have been produced in embryonated chicken eggs; this procedure is lacking with respect to efficient scalability, can involve the handling of live infectious virus, and cannot accommodate the need for hundreds of millions of doses per year. It is this global spread of highly pathogenic strains of influenza virus, such as the H5N1 subtype, that has provided the driving force toward the development of novel vaccines against pandemic influenza. Avian influenza strain H5N1, for example, has been identified as a potential pandemic threat by the World Health Organization (WHO), and over the past 12 years has spread from Asia to Europe and Africa. This strain of influenza virus in particular has been of global concern for the poultry industry. To block and control the spread of pandemic H5N1 a vaccination strategy is clearly required, and a procedure by which a safe and efficacious vaccine can be generated that is low in cost and easy to up-scale is urgently needed. It is clear, therefore, that plant-based technologies for the production of pandemic influenza vaccine could reduce both cost and response times to the emergence of new virus strains. The fact that plant-made vaccines are also easy to administer and do not require needles or refrigeration renders them compelling for further application to developing countries. This next section provides a brief overview of a number of the approaches that have been employed to produce vaccines against pandemic influenza virus in plants.

Over the last few years, several plant-based technologies have evolved showing promise for the large-scale production of influenza vaccine antigens. For example, Spitsin et al. (2009) generated in plants variants of the HA1 antigenic domain of the H5N1 avian influenza virus HA fused to His/c-myc tags or Fc antibody fragments which were derived from mice or humans. These plant-produced chimeric peptides were then purified and used for intramuscular immunization of mice. Although a serum humoral immune response was elicited in the mice, none of the plant-made HA1 preparations were able to induce virus-neutralizing antibodies. The presence of the Fc fragment, while assisting in increasing expression levels and simplifying purification, was unable to enhance the quality of the immune response. Another group designed a plant-derived vaccine against highly pathogenic avian influenza (HPAI) and tested it for its ability to protect poultry (Kalthoff et al., 2010). The results showed that this new vaccine can protect chickens against lethal challenge, further confirming the use of plant-derived vaccines for medical and veterinary purposes.

Kanagarajan et al. (2012) transiently expressed the haemagglutinin (HA0) of a poorly pathogenic strain of avian H7N7 influenza virus, isolated from a Swedish mallard duck, in tobacco plants using the cowpea mosaic virus (CPMV)-based vector (pEAQ-HT), along with the viral gene-silencing suppressor p19 from tomato bushy stunt virus. The authors recovered purified, N-glycosylated rHA0 protein at a concentration of 0.2 g purified protein kg^{-1} fresh weight of leaves. This purified rHA0 was able to exhibit both haemagglutination and haemagglutination inhibition activity, indicating that the protein maintained its native antigenicity and could be a potential source of vaccine antigen for poultry immunization.

Influenza virus antigens have also been expressed in plants in the form of virus-like particles (VLPs). As another example of the use of plant production platforms to produce an influenza vaccine, potato virus X (PVX) VLPs displaying the H-2D(b)-restricted epitope ASNENMETM of influenza A virus nucleoprotein (NP) were generated in plants. C57BL/6J mice were immunized and tested at different time intervals for splenocyte responses (Lico *et al.*, 2009). The plant-derived antigen was shown to be capable of activating ASNENMTEM-specific CD8+ T cells even in the absence of adjuvant.

Nemchinov and Natilla (2007) generated an expression system based on the coat protein (CP) of cucumber mosaic virus (CMV) that is placed under transcriptional control of a potato virus X (PVX)-based vector. The CMV CP that was expressed from this PVX-based expression vector system formed VLPs that were used to carry the 23 amino acid-long extracellular domain of the viral M2 protein (M2e) of influenza virus. Antibodies specific to the synthetic M2e epitope of influenza strain H5N1 reacted to the epitope on these VLPs.

Ravin *et al.* (2012) constructed a PVX vector that expresses an M2eHBc protein in tobacco plants. This construct consists of an extracellular domain of influenza virus M2 protein (M2e) fused to the hepatitis B core antigen (HBc). The chimeric protein was expressed to levels as great as 1–2% of the total soluble protein and was able to form VLPs with the M2e peptide displayed on the surface of the VLPs. Once purified, the VLPs exhibited a high level of protection against lethal influenza challenge in mice.

Babin *et al.* (2013) have engineered a novel adjuvant based on a nanoparticle of papaya mosaic virus (PapMV) CP, which has been assembled in a rod-like fashion around a RNA. A peptide antigen can be fused to a surface-exposed portion of the PapMV CP. This increases the immunogenicity of the epitope and thus provides a novel plant-based vaccine platform for infectious diseases such as influenza virus. The authors demonstrated that a cytotoxic T lymphocyte (CTL) epitope derived from the nucleocapsid of influenza virus (NP$_{147-155}$) could trigger a CTL response.

Matić *et al.* (2011) used VLPs based upon human papillomavirus 16 (HPV-16) L1 protein to act as a carrier of two foreign epitopes derived from the ectodomain of the M2 protein (M2e) of influenza A virus. These constructs were expressed from the cowpea mosaic virus expression vector, demonstrating the ability of human papillomavirus particles to act as effective carriers of vaccine epitopes, even within a plant background.

Another technology that produces VLPs carrying influenza virus antigen for prophylactic host inoculation in *Nicotiana benthiama* (D'Aoust *et al.*, 2008) has been developed by Medicago, Inc. VLPs expressing a lipid-anchored recombinant HA were shown to induce a fully protective immune response against lethal viral challenge in mice and ferrets. This H5-VLP pandemic vaccine was examined further in a Phase 1 clinical trial consisting of 48 healthy volunteers who received Medicago's vaccine at doses of 5, 10 or 20 μg, or a placebo. This trial demonstrated that the vaccine was safe and well tolerated. Furthermore, 81% of the immunized subjects were able to develop an immune response against the H5N1 virus (http://www.idri.org/press-04-17-13.php). A Phase 2 clinical trial with approximately 255 healthy volunteers is now being conducted (http://www.drugs.com/clinical_trials/medicago-reports-positive-phase-ii-interim-results-avian-flu-pandemic-vaccine-11104.html). Medicago has offered the potential for substantial up-scaling capacity, which can greatly outstrip the current pace of influenza vaccine production at a significantly lower cost (Kalthoff *et al.*, 2010; Landry *et al.*, 2010; Vezina *et al.*, 2011). This production system is capable of generating a vaccine within 3 weeks of the release of influenza strain sequence information.

Much progress concerning the generation of a plant-made flu vaccine has been made by researchers at the Fraunhofer USA Center for Molecular Biotechnology. In this instance, a tobacco mosaic virus-based 'launch' vector has been used to express vaccine proteins at high levels in plants. This vector consists of TMV sequences harboured within an *Agrobacterium* binary vector. The binary vector enables multiple single-stranded DNA copies of the virus construct, containing the vaccine epitope, to be released into the plant cell (Musiychuk *et al.*, 2007). For example, Shoji *et al.* (2009) expressed haemagglutinin from the A/Indonesia/05/05 strain of H5N1 influenza virus using this transient plant virus-based expression strategy. Mice and ferrets that were administered with plant-derived haemagglutinin were able to elicit an immune response that could protect the animals against virus challenge. In 2009, Shoji *et al.* also engineered and produced recombinant haemagglutinin (HA) from A/Bar-headed Goose/Qinghai/1A/05 (clade 2.2) and A/Anhui/1/2005 (clade 2.3) in tobacco and were able to produce serum haemagglutination inhibition (HI) and virus neutralization (VN) antibodies in mice that were immunized with the plant-made antigen.

The Fraunhofer research group has used a plant viral vector-based transient expression system to examine the immune response to mice who were vaccinated intranasally or intramuscularly with plant-derived influenza H5N1 (A/Anhui/1/05) antigen, either alone or formulated with the adjuvant *bis*-(3′,5′)-cyclic dimeric guanosine monophosphate (c-di-GMP) (Madhun *et al.*, 2011). They found that used intramuscularly, this adjuvant did not enhance the immune response to plant-derived influenza H5 antigen. Intranasal application of c-di-GMP-adjuvanted vaccine was able to induce strong mucosal and systemic humoral immune responses, including a high number of multifunctional Th1 CD4(+) cells. This group also expressed the haemagglutinin antigen of H1N1 influenza A strain (HAC1) that was derived from the 2009 pandemic H1N1 (pdmH1N1) virus, in *Nicotiana benthamiana* plants (Iyer *et al.*, 2012; Jul-Larsen *et al.*, 2012). The researchers were interested in testing the potential of this plant-derived vaccine to be commercially viable, and they conducted a series of studies that examined its potential in an animal model as well as in a clinical trial with human volunteers (Yusibov *et al.*, 2011). Using the aluminium salt adjuvant (Alhydrogel®), mouse immunogenicity studies demonstrated that high levels of serum immune responses (haemagglutination-inhibiting antibody titres) with a 100% seropositive rate, could be induced by HAC1 in the presence of this adjuvant. Furthermore, volunteers who were vaccinated with this vaccine plus adjuvant were used to demonstrate that by 7 days post-vaccination, the vaccine was able to be recognized by specific serum antibodies and antibody-secreting cells. The research group found a peak of serum HAC1 specific antibody response between days 14 and 21 post-vaccination, and demonstrated that the HAC1 antigen could be recognized by both T cells and B cells. Chichester *et al.* (2012) went on to conduct a Phase 1 randomized, double-blind, placebo-controlled clinical trial using three escalating dose levels of this vaccine, HAI-05, (15, 45 and 90 µg) adjuvanted with or without Alhydrogel®. No adverse reactions were reported when the vaccine was administered intramuscularly in two injections 3 weeks apart to healthy individuals, and the immune response tended to be variable, with the largest immune response observed in individuals immunized with the highest concentration of vaccine. The research group went a step further and generated a trimeric HA (tHA-BC) vaccine by introducing a trimerization motif from a heterologous protein into the HA sequence itself, to improve immunogenicity and stability of the antigen (Shoji *et al.*, 2013). The scientists found that this new construct of HA was able to protect mice against lethal viral challenge, and at a much lower dose than the previous, monomeric HA vaccine construct.

The fact that both the USA and Canada have initiated such large-scale manufacturing strategies for an influenza vaccine indicates that plant-made vaccines that target influenza virus are clearly up and coming technologies that could realistically be applied to developed and developing countries alike. The high cost of dealing with seasonal flu shots and the economic losses due to sick employees provide a sound basis for furthering the technology to the commercial level, and this is currently being reflected by the series of clinical trials that are proceeding at present. The results of these extensive and exhaustive clinical trials can only be beneficial to the world's rural poor and other population sets who also will be exposed to pandemic flu but lack access to the limited vaccine strategies offered today. In the case of some pandemic infectious diseases such as the H5N1 influenza virus, developed countries may run out of stockpiles quickly and would not be able to provide extra flu vaccines to developing countries. Ed Rybicki's group in South Africa has tested out the feasibility of generating plant-made vaccines for influenza virus, as a possibility for developing countries to supply their populace with their own 'homegrown' vaccine if none was available from other sources (Mortimer *et al.*, 2012). The researchers generated a plant-made vaccine for H5N1 using both transient assays as well as stably expressed transgenic plants. The vaccine consisted of a full length form of HA (H5) or a truncated form that lacked the transmembrane domain (H5tr). Both versions of the haemagglutinin were expressed at high levels in plants and were functional in haemagglutination and haemagglutination inhibition tests. Chickens and mice immunized with either form of the vaccine were able to elicit HA-specific antibody responses, providing proof-of-concept to the idea of a rapid response strategy by a developing country to the dangers of a potential pandemic.

10.2 Conclusions

Plant-made vaccines and therapeutic agents have been slowly but steadily gaining momentum as products that will take their place in the international marketplace. At present, a number of plant-made vaccine candidates are currently completing Phase III clinical trials and several that have been awarded approval are in a position to make their presence felt in critical areas of the pharmaceutical industry (Paul *et al.*, 2013). The example of the highly scalable, inexpensive and efficacious vaccine for pandemic flu provides a prominent example of how this technology will be compelling enough in today's competitive environment to readily find a commercial niche. With this in mind, it is not a far stretch of the imagination to envision how countries of the developing world can use this technology to be ensured more access to much-needed pharmaceuticals. While it is clear that pharmaceutical companies do not make it a priority to target infectious diseases that affect the developing world as they cannot expect an adequate return on investment, the strong distribution of patents available for plant-made pharmaceuticals derived from academic and publicly funded research institutes will provide accessibility to research and acquired knowledge for developing countries (Hefferon, 2013). Future action steps could be to provide feasible scenarios by which these vaccines could be produced locally and administered to the populace of the developing country under consideration.

While plant-made vaccines continue to hold promise, more issues need to be taken into account before this technology will make a positive statement on global health. It is not easy to predict how the introduction of these vaccines will play out with respect to existing multinational pharmaceutical companies and current healthcare providers for developing countries. The presence of current non-governmental organizations that specialize in providing conventional vaccines for the rural poor in developing countries can provide a reasonable and realistic template for how this can be realized. This book has offered compelling examples as to why plant-made pharmaceuticals should be part of the solution for those across the world who still lack access to even some of the most basic of medicines. Improvements in health will lead to enriching the overall quality of life for the world's poor, and this in turn would eventually expand workforce ability and, in the long run, play a role in facilitating an economic overhaul for these countries. Plant-made pharmaceuticals, therefore, could offer a helping hand by enabling people of the developing world to achieve and deliver their full potential as citizens of the global community.

References

Babin, C., Majeau, N. and Leclerc, D. (2013) Engineering of papaya mosaic virus (PapMV) nanoparticles with a CTL epitope derived from influenza NP. *Journal of Nanobiotechnology* 11, 10.

Brammer, L., Blanton, L., Epperson, S., Mustaquim, D., Bishop, A., Kniss, K., Dhara, R., Nowell, M., Kamimoto, L. and Finelli, L. (2011) Surveillance for influenza during the 2009 influenza A (H1N1) pandemic – United States, April 2009–March 2010. *Clinical Infectious Disease* 52, S27–35.

Chichester, J.A., Jones, R.M., Green, B.J., Stow, M., Miao, F., Moonsammy, G., Streatfield, S.J. and Yusibov, V. (2012) Safety and immunogenicity of a plant-produced recombinant hemagglutinin-based influenza vaccine (HAI-05) derived from A/Indonesia/05/2005 (H5N1) influenza virus: a phase 1 randomized, double-blind, placebo-controlled, dose-escalation study in healthy adults. *Viruses* 4(11), 3227–3244.

D'Aoust, M.-A., Lavoie, P.O., Couture, M.M., Trépanier, S., Guay, J.M., Dargis, M., Mongrand, S., Landry, N., Ward, B.J. and Vézina, L.P. (2008) Influenza virus-like particles produced by transient expression in *Nicotiana benthamiana* induce a protective immune response against a lethal viral challenge in mice. *Plant Biotechnology Journal* 6, 930–940.

Hefferon, K. (2013) Plant-derived pharmaceuticals for the developing world. *Biotechnology Journal* 8(10),1193–202.

Iyer, V., Liyanage, M.R., Shoji, Y., Chichester, J.A., Jones, R.M., Yusibov, V., Joshi, S.B. and Middaugh, C.R. (2012) Formulation development of a plant-derived H1N1 influenza vaccine containing purified recombinant hemagglutinin antigen. *Human Vaccine Immunotherapy* 8(4), 453–464.

Jul-Larsen, Å., Madhun, A.S., Brokstad, K.A., Montomoli, E., Yusibov, V. and Cox, R.J. (2012) The human potential of a recombinant pandemic influenza vaccine produced in tobacco plants. *Human Vaccine Immunotherapy* 8(5), 653–661.

Kalthoff, D., Giritch, A., Geisler, K., Bettmann, U., Klimyuk, V., Hehnen, H.R., Gleba, Y. and Beer, M. (2010) Immunization with plant-expressed hemagglutinin protects chickens from lethal highly pathogenic avian influenza virus H5N1 challenge infection. *Journal of Virology* 84(22), 12002–12010.

Kanagarajan, S., Tolf, C., Lundgren, A., Waldenström, J. and Brodelius, P.E. (2012) Transient expression of hemagglutinin antigen from low pathogenic avian influenza A (H7N7) in *Nicotiana benthamiana*. *PLoS One* 7(3), e33010.

Landry, N., Ward, B.J., Trépanier, S., Montomoli, E., Dargis, M., Lapini, G. and Vézina, L.P. (2010) Preclinical and clinical development of plant-made virus-like particle vaccine against avian H5N1 influenza. *PLoS One* 5, e15559.

Lico, C., Mancini, C., Italiani, P., Betti, C., Boraschi, D., Benvenuto, E. and Baschieri, S. (2009) Plant-produced potato virus X chimeric particles displaying an influenza virus-derived peptide activate specific CD8+ T cells in mice. *Vaccine* 27(37), 5069–5076.

Madhun, A.S., Haaheim, L.R., Nøstbakken, J.K., Ebensen, T., Chichester, J., Yusibov, V., Guzman, C.A. and Cox, R.J. (2011) Intranasal c-di-GMP-adjuvanted plant-derived H5 influenza vaccine induces multifunctional Th1 CD4+ cells and strong mucosal and systemic antibody responses in mice. *Vaccine* 29(31), 4973–4982.

Matić, S., Rinaldi, R., Masenga, V. and Noris, E. (2011) Efficient production of chimeric human papillomavirus 16 L1 protein bearing the M2e influenza epitope in *Nicotiana benthamiana* plants. *BMC Biotechnology* 11, 106.

Mortimer, E., Maclean, J.M., Mbewana, S., Buys, A., Williamson, A.L., Hitzeroth, I.I. and Rybicki, E.P. (2012) Setting up a platform for plant-based influenza virus vaccine production in South Africa. *BMC Biotechnology* 12, 14.

Musiychuk, K., Stephenson, N., Bi, H., Farrance, C.E., Orozovic, G., Brodelius, M., Brodelius, P., Horsey, A., Uqulava, N., Shamloul, A.M., Mett, V., Rabindran, S., Streatfield, S.J. and Yusibov, B. (2007) A launch vector for the production of vaccine antigens in plants. *Influenza Other Respiratory Viruses* 1(1), 19–25.

Nemchinov, L.G. and Natilla, A. (2007) Transient expression of the ectodomain of matrix protein 2 (M2e) of avian influenza A virus in plants. *Protein Expression and Purification* 56(2), 153–159.

Paul, M.J., The, A.Y., Twyman, R.M. and Ma, J.K. (2013) Target product selection – where can molecular pharming make the difference? *Current Pharmaceutical Diseases* 19(31), 5478–5485.

Perdue, M.L., Arnold, F., Li, S., Donabedian, A., Cioce, V., Warf, T. and Huebner, R. (2011) The future of cell culture-based influenza vaccine production. *Expert Review of Vaccines* 10, 1183–1194.

Ravin, N.V., Kotlyarov, R.Y., Mardanova, E.S., Kuprianov, V.V., Migunov, A.I., Stepanova, L.A., Tsybalova, L.M., Kiselev, O.I. and Skryabin, K.G. (2012) Plant-produced recombinant influenza vaccine based on virus-like HBc particles carrying an extracellular domain of M2 protein. *Biochemistry (Moscow)* 77(1), 33–40.

Shoji, Y., Chichester, J.A., Bi, H., Musiychuk, K., de la Rosa, P., Goldschmidt, L., Horsey, A., Uqulava, N., Palmer, G.A., Mett, V. and Yusibov, V. (2008) Plant expressed HA as a seasonal influenza vaccine candidate. *Vaccine* 26, 2930-2934.

Shoji, Y., Farrance, C.E., Bi, H., Shamloul, M., Green, B., Manceva, S., Rhee, A., Ugulava, N., Roy, G., Musiychuk, K., Chichester, J.A., Mett, V. and Yusibov, V. (2009) Immunogenicity of hemagglutinin from A/Bar-headed Goose/Qinghai/1A/05 and A/Anhui/1/05 strains of H5N1 influenza viruses produced in *Nicotiana benthamiana* plants. *Vaccine* 27(25–26), 3467–3470.

Shoji, Y., Jones, R.M., Mett, V., Chichester, J.A., Musiychuk, K., Sun, X., Tumpey, T.M., Green, B.J., Shamloul, M., Norikane, J., Bi, H., Hartman, C.E., Bottone, C., Stewart, M., Streatfield, S.J. and Yusibov, V. (2013) A plant-produced H1N1 trimeric hemagglutinin protects mice from a lethal influenza virus challenge. *Human Vaccine Immunotherapy* 9(3).

Spitsin, S., Andrianov, V., Pogrebnyak, N., Smirnov, Y., Borisjuk, N., Portocarrero, C., Veguilla, V., Koprowski, H. and Golovkin, M. (2009) Immunological assessment of plant-derived avian flu H5/HA1 variants. *Vaccine* 27(9), 1289–1292.

Vezina, L.-P., D'Aoust, M.A., Landry, N., Couture, M.J., Charland, N., Barbeau, B. and Sheldon, A.J. (2011) Plants as an innovative and accelerated vaccine-manufacturing solution. *BioPharmaceutical International Supplementary* 24, s27–30.

WHO (2010) *Animal and Pandemic Influenza*. Fifth Global Progress Report, July 2010.

WHO (2013) World: percentage of specimen tested positive for influenza – status as of week 52 (22–28 December 2013). Global Health Observatory, World Health Organization. Available at: http:// gamapserver.who.int/mapLibrary/Files/Maps/Global_influenzapositive_FluTransmissionZones_ week52_2013.jpg (accessed 22 September 2014).

Yusibov, V., Streatfield, S.J. and Kushnir, N. (2011) Clinical development of plant-produced recombinant pharmaceuticals: vaccines, antibodies and beyond. *Human Vaccines* 7(3), 313–321.

Index